Weather and Climate Resilience

DIRECTIONS IN DEVELOPMENT
Environment and Sustainable Development

Weather and Climate Resilience
Effective Preparedness through National Meteorological and Hydrological Services

David P. Rogers and Vladimir V. Tsirkunov

THE WORLD BANK
Washington, D.C.

© 2013 International Bank for Reconstruction and Development / The World Bank
1818 H Street NW, Washington DC 20433
Telephone: 202-473-1000; Internet: www.worldbank.org

Some rights reserved

1 2 3 4   16 15 14 13

This work is a product of the staff of The World Bank with external contributions. Note that The World Bank does not necessarily own each component of the content included in the work. The World Bank therefore does not warrant that the use of the content contained in the work will not infringe on the rights of third parties. The risk of claims resulting from such infringement rests solely with you.

The findings, interpretations, and conclusions expressed in this work do not necessarily reflect the views of The World Bank, its Board of Executive Directors, or the governments they represent. The World Bank does not guarantee the accuracy of the data included in this work. The boundaries, colors, denominations, and other information shown on any map in this work do not imply any judgment on the part of The World Bank concerning the legal status of any territory or the endorsement or acceptance of such boundaries.

Nothing herein shall constitute or be considered to be a limitation upon or waiver of the privileges and immunities of The World Bank, all of which are specifically reserved.

Rights and Permissions

This work is available under the Creative Commons Attribution 3.0 Unported license (CC BY 3.0) http://creativecommons.org/licenses/by/3.0. Under the Creative Commons Attribution license, you are free to copy, distribute, transmit, and adapt this work, including for commercial purposes, under the following conditions:

Attribution—Please cite the work as follows: Rogers, David P., and Vladimir V. Tsirkunov. 2013. *Weather and Climate Resilience: Effective Preparedness through National Meteorological and Hydrological Services.* Directions in Development. Washington, DC: World Bank. doi:10.1596/978-1-4648-0026-9. License: Creative Commons Attribution CC BY 3.0

Translations—If you create a translation of this work, please add the following disclaimer along with the attribution: *This translation was not created by The World Bank and should not be considered an official World Bank translation. The World Bank shall not be liable for any content or error in this translation.*

All queries on rights and licenses should be addressed to the Office of the Publisher, The World Bank, 1818 H Street NW, Washington, DC 20433, USA; fax: 202-522-2625; e-mail: pubrights@worldbank.org.

ISBN (paper): 978-1-4648-0026-9
ISBN (electronic): 978-1-4648-0027-6
DOI: 10.1596/978-1-4648-0026-9

Cover photo: ©NASA. Used with permission of NASA. Further permission required for reuse.
Cover design: Naylor Design.

Library of Congress Cataloging-in-Publication Data has been requested.

Contents

Foreword xi
Acknowledgments xiii
About the Authors xv
Abbreviations xvii

Chapter 1	Overview	1
	In This Chapter	1
	Introduction	1
	Why Are NMHSs Important?	3
	What Are the Obstacles to Better NMHSs?	4
	Key Principles for Modernizing NMHSs	6
	Notes	10
	References	11
Chapter 2	Coping with Weather, Climate, and Water Hazards	13
	In This Chapter	13
	Introduction	13
	A Snapshot of the Impact of Natural Disasters	14
	Warning Systems	16
	Forecasting Systems	18
	Arresting the Decline of NMHSs	19
	What the World Bank Can Do	25
	Notes	27
	References	28
Chapter 3	National Meteorological and Hydrological Services	31
	In This Chapter	31
	Introduction	31
	The Global Weather, Climate, and Water Enterprise	32
	The Special Role of NMHSs	37
	A Snapshot of Hydrological Services	42
	Latest in Forecasting Operations	47
	Limiting Factors in Forecasting	52

	Best Practices in Service Delivery	53
	A New Focus on Training	56
	Creating New Partnerships	59
	Top Priorities for Improving NMHSs	62
	Notes	64
	References	67
Chapter 4	**Best Practices in Warning Systems**	**71**
	In This Chapter	71
	Introduction	71
	Effective Warning Systems	72
	Core Elements of a Warning System	74
	How Multihazard Warning Systems Work	76
	Lessons from Shanghai's Multihazard System	81
	Notes	85
	References	86
Chapter 5	**Financing, Operating Models, and Regulatory Frameworks**	**89**
	In This Chapter	89
	Introduction	89
	Organization of NMHSs	90
	Funding	90
	A Need for Appropriate Operating Models	93
	Demand Side: The Users	94
	Supply Side: The Providers	96
	How Economic Characteristics Fit In	97
	Operating Models for NMHSs Services	101
	Public-Private Partnerships	108
	Legal and Regulatory Frameworks	112
	Notes	117
	References	118
Chapter 6	**Guidance on Modernizing NMHSs**	**121**
	In This Chapter	121
	Introduction	121
	Modernizing Advanced Meteorological and Hydrological Services	122
	Modernizing NMHSs in Developing Countries	123
	World Bank Experience in NMHSs Modernization	125
	Recommendations for Designing and Implementing Modernization Projects	129
	A Changing Role for NMHSs	136
	Notes	141
	References	141

Boxes

2.1	A Guide to Key Terminology	14
2.2	The World Meteorological Organization	18
2.3	Shortfalls in the Ability of the Lao People's Democratic Republic to Monitor River Flows	21
2.4	Threats to the Free and Unrestricted Exchange of Data	23
2.5	The European INSPIRE Program	24
2.6	The World Bank's GFDRR Hydromet Program	26
3.1	The World Meteorological Organization's Structure	33
3.2	How Switzerland Supports the Global Climate Observing System	37
3.3	Technical Insight: Functions of the Global Data Processing and Forecasting System	38
3.4	Technical Insight: World Meteorological Organization Global Centers	39
3.5	Technical Insight: World Meteorological Organization Regional Specialized Centers	40
3.6	Technical Insight: Key Types of River Flow Measurements	43
3.7	Technical Insight: Ensuring the Usefulness of River Flow Data	45
3.8	Hydrological Monitoring in the United Kingdom	46
3.9	Technical Insight: How Weather and Climate Interact	48
3.10	Technical Insight: Using Radiosondes for Upper-Air Measurements	48
3.11	Technical Insight: Modern Forecasting Techniques	49
3.12	Technical Insight: What Is Nowcasting?	50
3.13	Technical Insight: The Uncertainties Surrounding Long-Range Forecasting	51
3.14	Kenyan Farmers Need Better Forecasts	55
3.15	Attributes of Service Delivery Defined by the World Meteorological Organization	56
3.16	Teaching Meteorologists How to Communicate with Economists and Sociologists	57
3.17	U.S. Efforts to Broaden Training in the Environmental Sciences	58
3.18	On-Site Delivery of Localized Forecasts at Heathrow Airport	60
3.19	Technical Insight: Oklahoma Mesonet	61
3.20	Technical Insight: Severe Weather Forecasting Demonstration Projects	63
4.1	Technical Insight: The Nature of Disaster Risks and Hazards	73
4.2	Benefits of Partnerships for Early-Warning Systems	74
4.3	Ranking the Priority of Weather Warning Services	75
4.4	Cross-Cutting Issues for People-Centered Early-Warning Systems	77
4.5	Drawing Lessons from Hurricane Sandy	84
5.1	Organization and Staffing of National Meteorological Services	91
5.2	Funding Levels and Budgetary Pressures	92

5.3	Benefits of National Meteorological and Hydrological Services	95
5.4	Technical Insight: Quantifying Socioeconomic Benefits	96
5.5	Model I Example: U.S. National Weather Service	102
5.6	Model II Example: MeteoSwiss	104
5.7	Model III Example: U.K. Met Office	106
5.8	Model IV Example: New Zealand MetService	107
5.9	Model V Privatized Company: Holland Weather Services	109
5.10	Technical Insight: Public-Private Partnership between MeteoSwiss and Mobiliar	111
5.11	Technical Insight: Critical Elements of a Meteorological or Hydrological Law	114
6.1	Poland's National Meteorological Service Overhaul	127
6.2	The Russian Federation's Roshydromet Modernization	128
6.3	Lessons Learned from the Hydromet Modernization Process	130
6.4	Technical Insight: Component 1: Institutional Strengthening, Capacity Building, and Implementation Support	131
6.5	Technical Insight: Component 2: Modernization of the Observation Infrastructure and Forecasting	132
6.6	Technical Insight: Component 3: Enhancement of the Service Delivery System	133
6.7	Technical Insight: How to Encourage the Preparation of NMHSs' Modernization Projects	134
6.8	Technical Insight: A Spotlight on Doppler Radar	135
6.9	Technical Insight: Types of Twinning Arrangements	137
6.10	Technical Insight: Nepal's Results Framework	139

Figures

2.1	The Global Rise of Natural Disasters, 1975–2011	15
2.2	Average Annual Damages Caused by Reported Natural Disasters, 1990–2011	16
3.1	Activities of a Typical National Meteorological Service	32
3.2	World Meteorological Organization Global Network	34
4.1	Elements of a People-Centered Early-Warning System	75
4.2	Operational Flow of the Shanghai Meteorological Service Forecasting and Warning System	79
4.3	Severe Weather Warning Signals in Shanghai	80
4.4	Public Weather Service Work Flow	81
5.1	Weather and Climate Services Value Chain	94
5.2	Five NMHSs' Operating Models	102
B5.6.1	The Swiss Federal Council's Four Circles Model	104
6.1	World Bank Investments in Hydromet, Fiscal Years 1995–2011	126

Photos

3.1	Multiple observation systems	35
TI3.7.1	Vegetation obstructing water-level measurements	45

3.2	River gauging housing located on a tributary of the Mekong River, near Ho Chi Minh City, Vietnam. A rain gauge, satellite communication antenna, and solar panels are located on the roof	46
3.3	Modern forecaster workstation capable of blending observation and model products. Department of Meteorology, Ministry of Water Resources and Meteorology, Cambodia	49
3.4	Public Weather Service delivery system, (Meteorological and Geophysical Agency, Indonesia)	54
4.1	Part of the Shanghai Meteorological Service's delivery platform for public weather services and multi-hazard warnings	78

Tables

2.1	Deterioration of Hydrometeorological Observation Networks in Central Asia	20
2.2	Assessment of Economic Efficacy of Hydromet Modernization, Europe and Central Asia, 2006–08	26
3.1	Adoption of More User-Friendly Forecast Language in Croatia	55
5.1	Advantages and Risks of Public-Private Partnership Projects	110
6.1	World Bank Hydromet Programs	127
6.2	World Bank NMHSs' Modernization Projects in Poorer Countries	129

Foreword

Every day brings a reminder of how weather, climate and water affect the lives and livelihoods of each and every one of us, whether we live in Washington, D.C., Nairobi, or Dhaka. As this publication goes to print, more than 5,000 people are reported missing or presumed dead after major flooding in Uttarakhand, India; while in other parts of the world it is a lack of water coupled with extreme temperatures—that is taking lives and destroying livelihoods.

Developing economies are particularly vulnerable because they are predominately poor, less resilient and prone to natural hazards—and also because many don't have the capacity to provide risk information to their own citizens or manage disasters effectively.

As the world's population continues to grow, more and more people will be exposed to hazards. Over the past 30 years there has been a 114 percent increase in the number of people living in flood-prone river basins, and a 192 percent increase in the number of people exposed to tropical cyclones. Many of these people live in poorly planned urban areas, further increasing their vulnerability and exposure.

One part of the solution is to invest in structural mitigation measures (such as dams and levees); another is to map the hazards, identify vulnerabilities, provide accurate risk information and build early-warning platforms to enable early action in the face of extreme weather. Even a few minutes' warning can save lives, and longer lead times can help people protect livelihoods.

Effective early warnings can make the difference between life and death. But, as this report shows, the National Meteorological and Hydrological Services (NMHSs), tasked with producing such warnings, often lack capacity and resources. NMHSs in more than 100 countries—mostly in Africa—are in dire need of modernization investments.

By providing an overview of the vital role of NMHSs and highlighting the issues they face in many countries, this report aims to increase government and development agencies' understanding of the value of NMHSs and the commitments required to maintain their operations. More than that, the report provides convincing arguments that adequate support for NMHSs can have significant return on investment.

Ms. Zoubida Allaoua
Director, Urban and Disaster Risk Management Department,
Sustainable Development Network, the World Bank Group

Acknowledgments

A number of people contributed to this study, and we would like to thank them all. The staff of the World Meteorological Organization (WMO) was very helpful in providing information for this study, which draws heavily on WMO programs and operational centers.

The main authors of this study are David P. Rogers and Vladimir V. Tsirkunov. Chapter 3 benefited from the input of Peter Chen, Haleh Kootval, and Alice Soares of the WMO and the report "NMS Data Availability Study for World Bank" by Mike Gray of the U.K. Met Office. Tang Xu, director of the Shanghai Meteorological Service, provided most of the ideas for chapter 4 on multihazard early-warning systems. Chapter 5, which discusses operating models, was coauthored with Cornelia Giger, Gabriela Seiz, Christian Plüss, Fabio Fontana, and Michelle Stalder of MeteoSwiss, with contributions from Reto Steiner, Andreas Lienhard, and Etienne Huber of the Center of Competence for Public Management at the University of Bern, Switzerland, and Helen Martin of the World Bank. Lucy Hancock provided valuable input to chapter 6.

We would also like to thank Marianne Fay, Saroj Jha, and Francis Ghesquiere of the World Bank and Global Facility for Disaster Reduction and Recovery for their support of this study, as well as Eric Fernandez, Stéphane Hallegatte, Nagaraja Harshadeep, Marcus Wijnen, and Javier Zuleta, for reviewing and improving on early versions of the manuscript.

The authors would like to thank Claudia Sadoff, task team leader for the Pilot Program for Climate Resilience (PPCR) in Nepal, and Lia Sieghart, task team leader for the PPCR in the Republic of Yemen, for the opportunity to test ideas and concepts through the development of modernization programs for the national meteorological and hydrological services in those countries. The authors would also like to thank Henrike Brecht for providing comments and suggestions on behalf of the East Asia and Pacific Region of the World Bank. The authors are grateful to the staff members of various meteorological and hydrological services who shared their experiences. These institutions include the Australian Bureau of Meteorology, the Department of Meteorology in Cambodia, the China Meteorological Administration, the Finnish Meteorological Institute, Météo France, the Hong Kong Observatory, Met Eireann in Ireland, Kyrgyzhydromet in the Kyrgyz Republic, the Department of Meteorology and Hydrology in the Lao People's Democratic Republic, the Department of Hydrology and Meteorology in Nepal, Roshydromet in the Russian Federation, MeteoSwiss, Tajikhydromet in Tajikistan, the U.K. Met Office, the U.S. National Weather Service, the National Hydrometeorological Service in Vietnam, and the Civil Aviation and Meteorological Authority–Yemen Meteorological Service.

The authors are also grateful to Marjorie-Anne Bromhead, John (Jack) Hayes, and John Zillman for suggestions to improve the text and for external reviews of the manuscript; and to Laura Wallace, the principal editor of the report.

This publication was made possible with the support of the Global Facility for Disaster Reduction and Recovery (GFDRR) and its partners.

MEMBERS

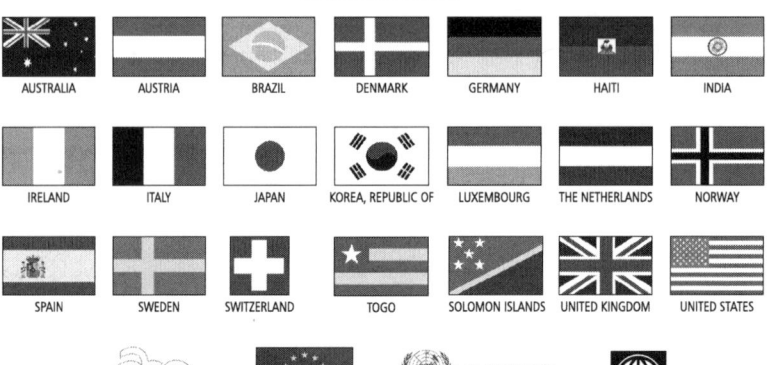

AFRICA, CARIBBEAN & PACIFIC (ACP) SECRETARIAT

EUROPEAN UNION

UNITED NATIONS OFFICE FOR DISASTER RISK REDUCTION

THE WORLD BANK

OBSERVERS

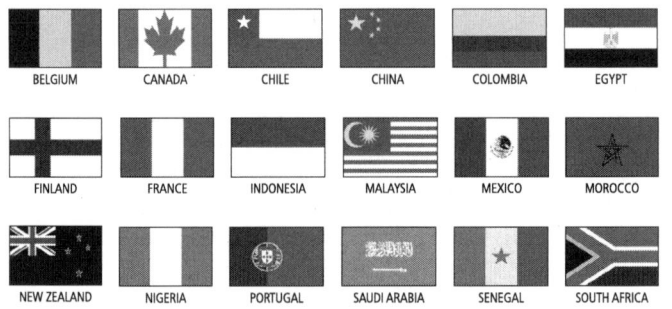

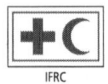
IFRC

ISLAMIC DEVELOPMENT BANK

UNITED NATIONS DEVELOPMENT PROGRAMME

Weather and Climate Resilience • http://dx.doi.org/10.1596/978-1-4648-0026-9

About the Authors

Dr. David P. Rogers is President of the Health and Climate Foundation (HCF), an international non-profit organization dedicated to finding solutions to climate related health problems and supporting partnerships between health and climate practitioners. Prior to founding HCF, Dr. Rogers held various appointments in government, the private sector and academia. These include Chief Executive of the U.K. Met Office; Vice President, Science Applications International Corporation; Director of the Office of Weather and Air Quality at the U.S. National Oceanic and Atmospheric Administration; Director of Physical Oceanography at Scripps Institution of Oceanography; and Associate Director of the California Space Institute, University of California, San Diego, USA.

Currently, Dr. Rogers is a consultant to the World Bank on modernizing National Meteorological and Hydrological Services.

Dr. Rogers has a Ph.D. (1983) from the University of Southampton and Bachelor of Science degree (1980) from the University of East Anglia, U.K. He has published extensively in the fields of oceanography, meteorology, climate, environment, and organizational development.

Dr. Vladimir V. Tsirkunov works as a Senior Environmental Engineer in the World Bank's Global Facility for Disaster Reduction and Recovery (GFDRR). Prior to joining the Bank in 1994, he was a Head of the Laboratory of the Supervision of the USSR (then Russian) System of Hydrochemical Monitoring and Assessment of Surface Water Quality. For the past 19 years, he has held progressively more responsible positions with the World Bank related to the preparation, appraisal, and operational supervision of environmental investments, technical assistance, and Global Environment Facility projects. Since 2003 he is developing new investment operations and analytical products supporting the improvement of weather, climate, and hydrological services, initially in Europe and Central Asia and later in East Asia, Africa, South Asia, and other regions. Since 2011, Dr. Tsirkunov has been leading GFDRR's Hydromet Program, which functions as a service center providing analytical, advisory, and implementation support for World Bank teams. Dr. Tsirkunov has a PhD (1985) from the Hydrochemical Institute and a bachelor of science degree (1977) from the Rostov State University, Russian Federation.

Abbreviations

AMDAR	Aircraft Meteorological Data Relay
ATC	air traffic control
AVHRR	Advanced Very High Resolution Radiometer
CMA	China Meteorological Administration
DHM	Department of Hydrology and Meteorology (Nepal)
EPS	ensemble prediction system
EUMETSAT	European Organisation for the Exploitation of Meteorological Satellites
GAW	Global Atmospheric Watch
GCOS	Global Climate Observing System
GDP	gross domestic product
GDPFS	Global Data-Processing and Forecasting System
GFDRR	Global Facility for Disaster Reduction and Recovery
GMES	Global Monitoring for Environment and Security (European Union)
GOS	Global Observing System
GPS	global positioning system
GTS	Global Telecommunication System
HWS	Holland Weather Services
HYCOS	Hydrological Cycle Observing System
ICT	information and communication technology
INSPIRE	Infrastructure for Spatial Information in the European Community
KNMI	Royal Netherlands Meteorological Institute
MHEWS	Multihazard Early-Warning System (Shanghai)
NCIC	National Climate Information Centre (United Kingdom)
NHSs	National Hydrological Services
NMC	National Meteorological Center
NMHSs	National Meteorological and Hydrological Services
NMSs	National Meteorological Services

NTSB	National Transportation Safety Board
NWP	numerical weather prediction
NWS	National Weather Service (United States)
O&M	operating and maintenance (costs)
PPCR	Pilot Program for Climate Resilience
PPP	public-private partnership
PR	permanent representative
PWS	Public Weather Service
R&D	research and development
RCCs	Regional Climate Centers
RSMC	Regional Specialized Meteorological Center
RTC	Regional Training Centers
SAR	special administrative region
SEVIRI	Spinning Enhanced Visible and Infrared Imager
SMS	Shanghai Meteorological Service
SOE	state-owned enterprise
SOP	standard operating procedure
SWFDP	Severe Weather Forecasting Demonstration Project
SWIC	Severe Weather Information Centre (Hong Kong SAR, China)
TCWC	Tropical Cyclone Warning Centers
THORPEX	The Observing System Research and Predictability Experiment
UNFCCC	United Nations Framework Convention on Climate Change
WCIDS	weather and climate information and disaster-support (systems)
WHYCOS	World Hydrological Cycle Observing System
WIGOS	WMO Integrated Global Observing System
WIS	WMO Information System
WMC	World Meteorological Center
WMO	World Meteorological Organization

CHAPTER 1

Overview

In This Chapter

Faced with the growing risks of weather and climate disasters to economic and social development, the global community needs to act quickly to strengthen National Meteorological and Hydrological Services (NMHSs). This strengthening should be done in a way that transforms weak agencies—especially in the developing world—into robust professional agencies capable of delivering the right information to the right people at the right time. Although the price tag of modernizing and sustaining NMHSs will be considerable, the rewards for the country and its citizens will be much higher. The World Bank, working closely with the World Meteorological Organization (WMO) and other development partners, can help countries navigate this complex task in a timely, efficient manner.

Introduction

The importance of weather, climate, and water[1] information is rising because of the need to serve more elaborate societal needs, minimize growing economic losses, and help countries adapt to climate change. Weather, climate, and water affect societies and economies through extreme events, such as tropical cyclones, floods, high winds, storm surges, and prolonged droughts, and through high-impact weather and climate events that affect demand for electricity and production capacity, planting and harvesting dates, management of construction, transportation networks and inventories, and human health.

The key players are the NMHSs,[2] which are the backbone of the global weather and climate enterprise. By international agreement under the auspices of the WMO,[3] they are the government's authoritative source of weather, climate, and water information, providing timely input to emergency managers, national and local administrations, the public, and critical economic sectors.

NMHSs are a small but important public sector—with budgets of usually about 0.01–0.05 percent of national gross domestic product and total annual public funding globally of more than US$15 billion. The problem is that their capacity has become so degraded in many regions over the past 15–20 years—primarily owing to underfunding, low visibility, economic reforms, and in some instances military conflict—that they are now inadequate. As a result, globally, NMHSs in more than 100 countries—over half of which are in Africa—need to be modernized.

How much will modernization cost? A conservative estimate of high-priority modernization investment needs in developing countries exceeds US$1.5 billion to US$2.0 billion. In addition, a minimum of US$400 million to US$500 million per year will be needed to support operations of the modernized systems (staff costs plus operating and maintenance costs). These recurrent costs should be covered by national governments, but few are ready to do so. Moreover, the amount of international support for the NMHSs is significantly below what is needed just for the high-priority items.

Complicating matters is that internationally supported NMHSs modernization efforts in the developing world have achieved only limited success so far, owing to the following reasons:

- A lack of government and development agencies' understanding of the value of the NMHSs and a lack of commitment to maintain their operations
- A preoccupation with project time-scale installation of hardware without adequate provision for training, ongoing maintenance, consumables, and other continuing technical support
- A multiplicity of uncoordinated projects from different donors, each with its own assistance policies, objectives, and equipment suppliers, without sufficient regard to the individual NMHSs' needs, circumstances, and priorities
- The technical complexity of the projects

What can be done to improve the track record of modernization efforts and help policy makers realize the urgent need to overhaul NMHSs? To help answer this question, we undertook a study that analyzes the overall global meteorological and hydrological system, its importance for the effectiveness of NMHSs, the obstacles to modernization, and possible desirable operating models.

Our study combines desktop analyses based on existing documentation with expert opinion and the experience of several leading NMHSs—the U.K. Met Office, MeteoSwiss, and the China Meteorological Administration. It synthesizes recent experiences of the World Bank and Global Facility for Disaster Reduction and Recovery (GFDRR), the WMO, and other development partners. It also builds on key recent World Bank strategic documents, including (a) *World Development Report 2010: Development and Climate Change* (World Bank 2009); (b) the 2012–22 environmental strategy, which argues for green, clean, resilient development (World Bank 2012c); and (c) the 2012 green growth strategy, which articulates the case for both inclusive economic growth and environmental sustainability (World Bank 2012a).

The report underscores the urgent need to strengthen NMHSs, especially those in developing countries, and provides cost-benefit estimates of the return that countries can hope to achieve. It also offers a recommended approach that has been tested and implemented in Europe, in Central and South Asia, and in countries in other regions. And it underscores the significance of international collaboration to access data, knowledge, and know-how of the large-scale atmospheric and oceanic conditions that drive the global weather patterns that affect individual countries.

It has been conservatively estimated that upgrading all hydrometeorological information production and early-warning capacity in developing countries would save an average of 23,000 lives annually and would provide between US$3 billion and US$30 billion per year in additional economic benefits related to disaster reduction (Hallegatte 2012).

Why Are NMHSs Important?

Weather, Climate, and Water Hazards

In recent years—thanks largely to advances in weather forecasting and risk assessments—people have been better prepared for natural disasters. Despite an increase in the number of disasters and people affected since 1980, the number of people killed has not risen significantly. A huge concern is that not only will the number of people affected and the number of disasters continue to rise but also the number of people killed will increase if governments and other stakeholders do not intervene. The reasons are many:

- An increasing number of people and assets are located in areas of high risk.
- Developing countries will continue to be exposed to frequent and extreme weather events as climate change exacerbates these extremes.
- The world's population continues to explode.
- The urbanization trend continues, with more people living in cities than ever before.
- Weather- and climate-sensitive diseases claim more than 1 million lives each year; most are children under five years of age in developing countries.

Between 1970 and 2010, natural hazards killed about 3.3 million people (World Bank 2010). They also took a huge financial toll on human well-being. In 2011, about 206 million people were afflicted by natural disasters, and the economic impact was US$366 billion (UNISDR 2011). During a longer period, between 1980 and 2011, the total estimated financial cost from floods, droughts, and storms was more than US$3.5 trillion (Munich Re 2012).

Weather, Climate, and Water Forecasts

The NMHSs make a significant contribution to safety, security, and economic well-being by observing, forecasting, and warning of pending weather, climate, and water threats. However, this contribution is rarely quantified, which often results in undervaluing the vital role NMHSs play in a country's capacity to cope with meteorological and hydrological hazards. Also severely undervalued are the economic benefits of accurate weather, climate, and water information to increase productivity and avoid losses.

Accurate forecasting depends on a network of global, regional, and national remote and in situ observations of the atmosphere, oceans, and land that are conducted by NMHSs and their partners. These observations are assimilated by a network of global and regional forecast centers, which have differentiated responsibilities for the production of global, regional, and national products. This system ensures that large-scale numerical predictions—which are needed for a good national forecast but require enormous computing power—are created cost-effectively by a few NMHSs and supporting organizations on behalf of all Members of the WMO.

Alone, no nation would be able to provide the meteorological and hydrological services necessary to meet the essential needs of its citizens. But as WMO Members, countries agree on data-sharing arrangements, establish operational guidelines, implement best practices, and develop and use training opportunities. This international cooperation, however, depends on the continued investment of advanced countries in developing and supporting meteorological satellites, major computing facilities, and research and development. It also depends on regional investment in adapting global products for regional and national application. And it depends on national investment in maintaining NMHSs' observation networks and tailoring services to the needs of the population and specific economic sectors.

What Are the Obstacles to Better NMHSs?

Lack of Capacity

Despite their importance, many NMHSs in developing countries lack the capacity to provide even a basic level of services. The massive underfunding of NMHSs has led to (a) a deterioration of meteorological and hydrological observation networks and outdated technology, (b) a lack of modern equipment and forecasting methods, (c) poor quality of services, (d) insufficient support for research and development, and (e) an erosion of the workforce (resulting in a lack of trained specialists). As a result, substantial human and financial losses have occurred, which could have been avoided if weather and water agencies were more developed. Climate-resilient development requires stronger institutions and a higher level of observation, forecasting, and service delivery capacity. In addition, successful adaptation to the existing and future weather and climate variability is impossible without reliable and well-functioning NMHSs.

The problem is not limited to NMHSs alone. The WMO regional, specialized, and global centers should play a significant role in helping countries reach a high level of service. However, investment here is also limited, and the on-demand guidance from a WMO regional center that could be available to a country is often not. Although a national focus is primary, the benefit of international cooperation and collaboration must also be considered. Synergy between the different levels ensures that national data are available to improve model output at regional and global centers. The high value-added segment of the production chain with regard to numerical weather prediction and space-based observations is at the global level. At present, it is assumed that developed economies will continue to support this segment. But this assumption is becoming increasingly uncertain.

Public Financing of NMHSs

Nearly all NMHSs started as public service institutions that were funded exclusively by the government (Zillman 2005a).[4] But at the end of the 20th century, a series of economic policy shifts occurred in both developed and developing countries, resulting in public policies that are less supportive of this concept (Zillman 2005b). There has been a growing expectation among many governments that the public sector should raise revenue from the sale of services to other government departments and to the public. This view presents serious threats to the continued free and unrestricted exchange of information that the WMO has always pursued (Jean et al. 1999; Zillman 1999).

In response, institutional frameworks have been adjusted to give NMHSs more flexibility to generate revenue and to use that revenue to expand and improve services. However, as claims on limited tax revenue have increased, greater emphasis has been put on applying the principle that the user should pay for government services (Freebairn and Zillman 2002). Government budget allocations are offset by revenue generated from the provision of paid services. In the worst case, countries with very weak NMHSs resort to the practice of restricting public access to and selling data.

This commercial approach appears to result from a failure to understand the natural monopoly component of this type of information—which requires large fixed investments with a quasi-zero marginal distribution cost and thus should be available for free. Without the large fixed public (NMHSs) investments, it is impossible to sustain the observation networks, except for geographically constrained and specialized applications. Without these observations, basic forecasts are limited and value-added services to end users are not possible. This situation undermines civil protection, food security, water resource management, energy and emergency management, and economic development.

Choosing the Right Operating Model

Many governments have sought ways to reduce the cost of providing weather and hydrological services. A common approach has been to transform NMHSs from departments into agencies or state-owned companies. One reason is that entities that operate at arm's length from the central government may improve efficiency because the underlying structure allows specialization in service provision, with a stronger focus on customers' needs and increased competitiveness. Another reason is that a higher degree of flexibility and customer friendliness may increase effectiveness. A third reason is that decision-making processes may improve because of a lower degree of political influence and the greater inclusion of specialized professionals into the process. However, not all types of public services and goods are suitable for outsourcing.[5]

The five common operating models for NMHSs are (a) departmental unit, (b) agency, (c) public body, (d) state-owned enterprise, and (e) privatized company. These operating models range from no autonomy to full autonomy. But given the NMHSs' natural monopoly on observations[6] and primary responsibility to the public and government, full privatization does not appear to be a real option for NMHSs, but is reserved for spin-offs of public sector bodies that operate entirely within the commercial realm. Each of the other models can be further extended by including public-private partnerships, which can increase efficiency or improve service delivery. Which model a country chooses largely depends on the political, economic, and societal landscapes—although the most frequently used ones are departmental unit, agency, and public body.

Key Principles for Modernizing NMHSs

In response to the growing risk of meteorological and hydrological hazards, the study has identified six principles for improving NMHSs in developing countries:

Principle 1: Modernizing NMHSs in Developing Countries Is a High-Value Investment

Although the challenges in modernizing NMHSs are great, so too are the potential benefits to societies coping with meteorological and hydrological hazards and the risks posed by climate change. Globally, our capabilities are the best that they have ever been. Scientific and technological advances continue to improve numerical weather and climate prediction. We now have the scientific skills to provide reliable warnings of extreme events and day-to-day weather forecasts that are more accurate, specific, and timely than ever before—and these skills continue to improve. However, they are often limited to developed countries, because NMHSs in developing countries lack the infrastructure to transfer and use these technologies.

Unfortunately, many governments fail to understand the societal value of the information and services that NMHSs should provide as a public service. This part of the so-called poverty trap—namely, the existing poor status of NMHSs—prevents the production of valuable data and information. Governments see no reason, therefore, for investing in NMHSs. But without investment, there are no new products and services, a situation that is manifest in poor or nonexistent meteorological and hydrological warnings. Substantial, well-targeted, and long-term financial support and capacity building are needed to break this cycle, together with improved communication and advocacy campaigns.

One way to enhance government and broaden public understanding of what is at stake is to conduct socioeconomic studies that quantify the value of the public services resulting from NMHSs' strengthening. Such studies can also identify gaps in the current system and help prioritize elements of a modernization program. This process should be iterative so that stakeholders' expectations are realistic. Engaging all stakeholders, both internal and external to the NMHSs, is critical to the success of a modernization program. In Switzerland and the United States, studies show high economic returns from better NMHSs—with cost-benefit ratios of 1:4–1:6. And a recent World Bank study in Europe and Central Asia suggests cost-benefit ratios of 1:2–1:10 (Tsirkunov et al. 2007).

Principle 2: The Financing and Scope of Modernization Must Be Sufficient to Be Transformative

Financing and scope of modernization must be enough to change NMHSs with poor infrastructure, declining observation networks, and weak forecasting capability into public service organizations capable of delivering timely and useful information to mitigate weather, climate, and water risks to the public and sensitive economic sectors. New capabilities incur additional operating and maintenance costs, which governments must consider up front to ensure the sustainability of the modernization effort beyond the initial work program.

The appropriate operating models need to be recognized explicitly to ensure that the NMHSs meet their public service and international obligations. Governments need to recognize and support their NMHSs to protect lives, livelihoods, and property as a critical, publicly funded mission. Policies that may restrict the free and open exchange of meteorological and hydrological data should be avoided, and the public sector responsibilities of the NMHSs should be emphasized. Selecting an operating model goes hand in hand with establishing appropriate legislation to institutionalize the agreed mission.

Principle 3: Clear Legal and Regulatory Frameworks for Providing Weather, Climate, and Water Services Increase Effectiveness

Broad engagement across government departments, agencies, and other institutions is essential for success. To achieve success, countries need legal and regulatory frameworks for providing meteorological and hydrological warnings, as well as for delivering other weather, climate, and water services. Such frameworks will enable all stakeholders to understand their respective roles and responsibilities and to act accordingly. Coordination across government agencies is difficult, if not impossible, without it.

Principle 4: Large-Scale Modernization Programs Should Typically Include Three Components

These components are as follows:

- *Institutional strengthening, capacity building, and implementation support.* Strengthening NMHSs' legal and regulatory frameworks; improving their institutional performance as the main provider of weather, climate, and hydrological information for the country; building the capacity of personnel and management; ensuring operability of future networks; and supporting project implementation are all necessary to a large-scale modernization program.

- *Modernization of observation infrastructure and forecasting.* This component includes modernizing the NMHSs' observation networks and communications and information and communication technology (ICT) systems, improving the meteorological and hydrological forecasting systems, and refurbishing offices and facilities.

- *Enhancement of the service delivery system.* Such enhancement involves creating or strengthening the public weather services, climate services, and hydrological services and developing new information and value-added products for vulnerable communities and the main meteorology- and hydrology-dependent sectors. This component should include developing a national framework for climate services, considered within the context of the Global Framework for Climate Services (GFCS).

The World Bank's experience suggests that NMHSs need help to transform their operations. They need expert guidance throughout the modernization process, which can be provided by a general consultant or systems integrator or a group of individual consultants for smaller programs. In addition, pairing advanced NMHSs with less advanced NMHSs helps sustain staff training and provides operational guidance, especially in advance of extreme hydrometeorological hazards.

Principle 5: Modernization of NMHSs Should Be Considered within the Wider Regional and Global Contexts

It is important to understand which parts of the public meteorological infrastructure are best funded and operated at the local, national, regional, and global levels and to make investments accordingly. There is room for more efficient distribution of responsibilities among these levels. Technological developments make it possible to generate more useful products at regional and global levels, which can underpin the services that NMHSs provide at the country level.

WMO regional centers and specialized centers are an integral part of the information system. They provide NMHSs with operational guidance based on the products created by the global modeling centers. Strong regional and specialized centers can help sustain national modernization programs by supporting continuous technology infusion, thereby ensuring that the NMHSs are up to date. However, new financing mechanisms are needed to support the regional and global elements of the meteorological and hydrological system.

Principle 6: The World Bank and Development Partners Have a Vital Role in Strengthening NMHSs

The reason their role is so vital is simple: weather, climate, and water services are a key public good, and better resilience to climate variability and change is a key element of a broader sustainable development and green growth agenda. Since the mid-1980s, the World Bank has prepared and implemented more than 150 operations with some elements supporting NMHSs, but relatively few were aimed at modernizing the whole system. Rather, the investments were structured as small-scale activities within water resource management, agriculture, or emergency operations. The approach was often piecemeal, emphasizing efforts to patch up services by supplying individual sensors and partial systems, without a strong connection with the National Meteorological Services (NMSs) or users.

But since the mid-1990s, the focus has shifted toward development of a more holistic approach.[7] And today, most efforts involve modernizing entire NMHSs—through institutional strengthening, improving observation networks and forecasting, and strengthening service delivery. In collaboration with the WMO, the World Bank has an advisory role in helping to inform governments of the high societal and economic significance of weather, climate, and hydrological information and services and of the importance of making meteorological and hydrological agencies the center of this support. The World Bank is also helping NMHSs raise their profiles in their respective governments by using the results of economic assessments, cost-benefit analyses, and analytical work, along with identifying priority investment needs and facilitating financial support. The main instruments used by the World Bank are traditional lending and technical assistance projects—and it is investigating how to use the new financial instruments of climate adaptation and climate investment funds.

These modernization activities can be an integral part of larger projects in disaster reduction, water resource management, agricultural support, and public health improvement. The World Bank is also building stronger partnerships with the WMO and leading NMHSs. Currently, more than a dozen operations are in various stages of implementation, and many more are at the conceptual stage. The total cost of investments under preparation or implementation exceeds US$400 million.

To help strengthen NMHSs, the World Bank has created a support program—GFDRR Hydromet—which is part of the GFDRR. This program, formerly known as Strengthening Weather and Climate Information and Decision Support Systems, was launched in May 2011 as a vehicle to coordinate many of these activities.[8] It is also aimed at helping countries implement the Hyogo Framework for Action—a 10-year international plan adopted in 2005 to make the world safe from natural disasters.[9] The GFDRR Hydromet program includes (a) analytical support and knowledge management, (b) capacity building and technical assistance, and (c) development of the World Bank's portfolio and operations.

The World Bank is also beginning to scale up support through investment and development policy operations. For example, hydrometeorology forms a key pillar of all of the programs developed under the Pilot Program for Climate Resilience of the Climate Investment Funds. And countries are increasingly recognizing the importance of integrating better weather and climate service delivery into broader strategic development agendas.

Hydrometeorological programs are technically complex, but the World Bank has gained practical knowledge in modernizing NMHSs in middle-income and now low-income countries—although so far it has limited experience in the least developed countries. In 2012, new hydrometeorological modernization projects were being developed and implemented for Mexico, Mozambique, Nepal, the Russian Federation, Vietnam, the Republic of Yemen, Zambia, and other countries. In Nepal, the approach is one of the few examples of end-to-end modernization, focusing on institutional strengthening, modernization of the observation and forecasting systems, and service delivery.[10] In addition, a dedicated assessment of the current capacity of hydrological services will be the focus of a separate study that the World Bank is planning to undertake jointly with the WMO. The hope is that the World Bank's growing modernization experience can help improve the design of future programs and underpin the place of hydromet strengthening in broader sustainable development agendas.

Notes

1. The tag *weather, climate, and water* is used frequently instead of meteorology and hydrology. Meteorology is inclusive of weather and climate, and these terms are interchangeable. *Water* is the tag used to refer to hydrology and occasionally oceanography.
2. Collectively, National Meteorological Services and National Hydrological Services are referred to as NMHSs. The abbreviation NMS refers to a National Meteorological Service or, if hydrology and meteorology are combined in a single institution, a National Hydrometeorological Service.
3. The World Meteorological Organization (WMO) is a United Nations specialized agency with 191 member states and territories, each of which is represented by a permanent representative. The permanent representative is responsible for representing the national weather, climate, and hydrological community within the WMO and by convention is usually the director of the NMS or National Hydrological Service (NHS). The private sector and academia are represented within the WMO by permanent representatives or affiliated organizations.
4. According to Zillman (2005a, 225), among the key challenges is a need to establish "a rigorous and comprehensive economic framework for meteorology at both the national and international level"; a need to achieve "universal recognition of basic meteorological and related environmental data as a global public good"; and a need to build "robust and mutually supportive partnerships between the public, private, and academic sectors of meteorology at both the national and international level."
5. With respect to public goods and services, the term *outsourcing* tends to be used instead of *decentralization*.
6. Basic national meteorological and hydrological observation networks are usually funded entirely by government. They may be supplemented by small-scale networks for local application, but in general NMHSs have a natural monopoly in the supply of basic observations of the weather, climate, and water system.
7. The World Bank (2012b) summarizes the investments that have been made in NMHSs since 1995 and identifies some of the recurring issues that are being addressed in the current program.
8. For more information about the program, visit Global Facility for Disaster Reduction and Recovery's (GFDRR) website at https://www.gfdrr.org/hydrometservices.

9. For a description of the Hyogo Framework for Action, see the website of the United Nations International Strategy for Disaster Reduction at http://www.unisdr.org/we/coordinate/hfa.
10. Nepal's hydromet modernization program is called Building Resilience to Climate-Related Hazards and is funded by the Pilot Program for Climate Resilience.

References

Freebairn, John W., and John W. Zillman. 2002. "Funding Meteorological Services." *Meteorological Applications* 9 (1): 45–54.

Hallegatte, Stephane. 2012. "A Cost Effective Solution to Reduce Disaster Losses in Developing Countries: Hydrometeorological Services, Early Warning, and Evacuation." Policy Research Working Paper 6058, World Bank, Washington, DC.

Jean, Michel, Bruce Angle, David Grimes, and John Falkingham. 1999. "Structure and Evolution of National Meteorological and Hydrological Services: International Comparisons." *WMO Bulletin* 48 (2): 159–65.

Munich Re. 2012. *Topics: Annual Review—Natural Catastrophes, 2011*. Munich, Germany: Munich Re.

Tsirkunov, Vladimir, Sergey Ulatov, Marina Smetanina, and Alexander Korshunov. 2007. "Customizing Methods for Assessing Economic Benefits of Hydrometeorological Services and Modernization Programmes: Benchmarking and Sector-Specific Assessment." In *Elements for Life*, edited by Soobasschandra Chacowry. Geneva, Switzerland: World Meteorological Organization.

UNISDR (United Nations International Strategy for Disaster Reduction). 2011. "Preparing for RIO+20: Redefining Sustainable Development." Discussion Paper, UNISDR, Geneva, Switzerland, October 10.

World Bank. 2009. *World Development Report 2010: Development and Climate Change*. Washington, DC: World Bank.

———. 2010. *Natural Hazards, Unnatural Disasters: Effective Prevention through an Economic Lens*. Washington, DC: World Bank.

———. 2012a. *Inclusive Green Growth: The Pathway to Sustainable Development*. Washington, DC: World Bank.

———. 2012b. *Strengthening Weather and Climate Information and Decision Support Systems (WCIDS): The World Bank Portfolio 1995–2011*. Washington, DC: World Bank.

———. 2012c. *Toward a Green, Clean, and Resilient World for All: A World Bank Group Environmental Strategy 2012–2022*. Washington, DC: World Bank.

Zillman, John W. 1999. "The National Meteorological Service." *WMO Bulletin* 48 (2): 129–59.

———. 2005a. "The Challenges for Meteorology in the 21st Century." *WMO Bulletin* 54 (4): 224–29.

———. 2005b. "Real Time Data Requirements of National Meteorological Services (NMHSs) and Their Users." Paper presented at the National Oceanic and Atmospheric Administration's National Weather Service International Session on Addressing Data Acquisition Challenges, San Diego, CA, January 6–7.

CHAPTER 2

Coping with Weather, Climate, and Water Hazards

In This Chapter

Society is increasingly sensitive to meteorological and hydrological hazards. The costs of weather, climate, and water disasters are rising, but the capacity to effectively use weather, climate, and water information to reduce these costs is also greater. Investing in global hydrometeorological infrastructure—particularly in National Meteorological and Hydrological Services—is essential to cope with weather, climate, and water hazards. And the free and unrestricted sharing of international meteorological and hydrological data is critical to this task.

Introduction

The potential for natural hazards—such as intense floods, droughts, storms, and earthquakes—to undermine development is great and growing (DFID 2006; GFDRR 2012[1]; Guha-Sapir et al. 2012). These hazards hurt the economic performance of many developing countries and the lives and livelihoods of millions of poor people around the world (van Aalst 2006). Between 1970 and 2010, natural hazards killed about 3.3 million people (World Bank 2010). In addition, weather extremes put at risk investments in infrastructure, agriculture, human health, water resources, disaster management, and the environment. For example, the transportation infrastructure in Africa is crucial to bringing Africa out of poverty, but flooding incapacitates large parts of this network every year. Weather extremes also increase the vulnerability of people—particularly the poorest—when development needs trigger investment and human settlement in coastal zones, flood plains, arid areas, and other high-risk environments.

Fortunately, the capacity to effectively use weather, climate, and water information is also getting better. But this capacity will be realized only if countries step up their investments in a critical part of the national infrastructure—National Meteorological and Hydrological Services (NMHSs)—which have not been sufficiently well maintained in recent decades in both developed and

Box 2.1 A Guide to Key Terminology

- *Meteorology:* The scientific study of the Earth's atmosphere as it relates to short-term weather and longer-term climate variations.
- *Hydrology:* The scientific study of the Earth's water system.
- *Meteorological and hydrological hazards:* Flash floods, river floods, thunderstorms, tropical cyclones, and other extreme weather–related events, as well as slow-onset hazards, such as droughts.
- *Weather, climate, and water:* A tag used frequently instead of *meteorology and hydrology*. Meteorology is inclusive of weather and climate, and these terms are interchangeable. Water refers to hydrology and occasionally to oceanography. The term *meteorological* embraces both meteorological and climatological phenomena.
- *NHMSs:* An abbreviation that encompasses both National Meteorological Services (NMSs) and National Hydrological Services (NHSs). The abbreviation NMS also refers to a national hydro-meteorological service (if hydrology and meteorology are combined in a single institution).
- *Hydrometeorology* or *hydromet:* A term used when an organization has the combined responsibility for meteorology and hydrology.
- *Forecasting:* The application of science and technology to predict the state of the atmosphere for a given location on time-scales of hours to years. Forecasts are often referred to as *nowcasts* (from 0 to 6 hours), *very-short-range weather forecasts* (up to 12 hours), *short-range weather forecasts* (from 12 to 72 hours), *medium-range weather forecasts* (from 3 to 10 days), *extended-range weather forecasts* (from 10 to 30 days), and *long-range forecasts* (from 30 days to 2 years). There are also monthly, trimonthly, and seasonal outlooks (covering, for example, December to February, March to May, June to August, or September to November) and longer-term climate predictions (from years to centuries).

developing countries. This chapter looks at the big picture, starting with the state of natural disaster and early-warning systems; the challenges facing the NMHSs, including a threat to the free and unrestricted exchange of data; and the ways that the World Bank and global community can help. See box 2.1 for important terminology used throughout the book.

A Snapshot of the Impact of Natural Disasters

Heavy rain events, high winds, storm surges, flash floods, and river floods are meteorological and hydrological hazards. Disasters are deaths and damage that result from human acts of omission and commission (World Bank 2010). In other words, although exposure to hazards cannot be avoided, the potential ensuing disasters can be mitigated. In recent years—thanks largely to advances in weather forecasting and risk assessments—people have been better prepared for natural disasters. Despite the growing number of disasters and people affected since 1980, the number of people killed has not increased significantly (figure 2.1).[2]

A big worry is not only that the number of disasters and number of people affected will continue to rise but also that the number of people killed will

Figure 2.1 The Global Rise of Natural Disasters, 1975–2011

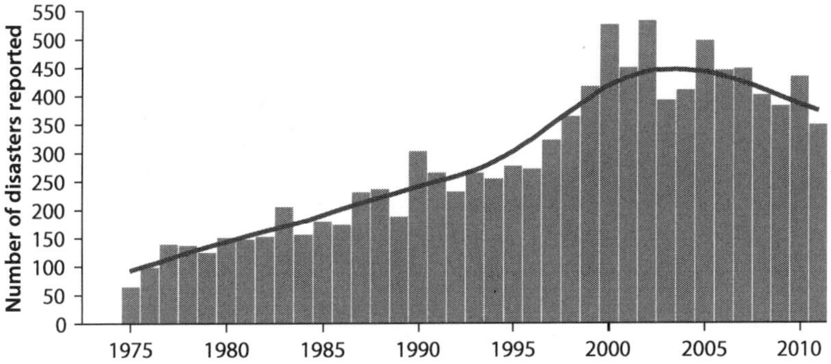

Source: EM-DAT, International Disaster Database, Centre for Research on the Epidemiology of Disasters, Brussels, http://www.emdat.be/natural-disasters-trends.

increase if governments and other stakeholders fail to intervene. One reason is that developing countries will continue to be exposed to frequent and extreme weather events as climate change exacerbates those extremes. A second reason is the continuing population explosion, with the world's population growing by 87 percent over the past 30 years, mainly in developing countries. A third reason is that, as a result, more people and assets are now located in areas of high risk. Over the same period, the proportion of the world's population living in flood-prone river basins has increased 114 percent, and the number of people living on cyclone-exposed coastlines has increased 192 percent (UNISDR 2011b). A fourth reason is urbanization, with more than half of the global population now living in cities for the first time in history. This trend has been most rapid in coastal areas, where more people are exposed to floods with limited structural protection, inadequate citywide drainage systems, and weak nonstructural mitigation measures. A fifth reason is that weather- and climate-sensitive diseases claim more than 1 million lives each year, mostly children under five years of age in developing countries.

As for the financial toll of natural hazards on human well-being, the United Nations International Strategy for Disaster Reduction estimates that in 2011 about 206 million people were afflicted by natural disasters, and the economic impact was US$366 billion. Although the Tohoku earthquake and tsunami[3] in Japan accounted for US$210 billion, the remainder was associated mostly with weather, climate, and hydrological events, which accounted for 89 percent of annual disasters. For example, flooding in Thailand from August to December caused US$40 billion, or 12.7 percent of gross domestic product (GDP), in economic damages (Guha-Sapir et al. 2012). During a longer stretch of time, between 1980 and 2011, the total estimated financial cost from floods, droughts, and storms was over US$3.5 trillion (Munich Re 2012).

Which countries bear the highest burdens? In absolute terms, the financial impact is largest in the richest countries. Since 1980, the risk of economic loss owing to floods in member countries of the Organisation for Economic

Figure 2.2 Average Annual Damages Caused by Reported Natural Disasters, 1990–2011

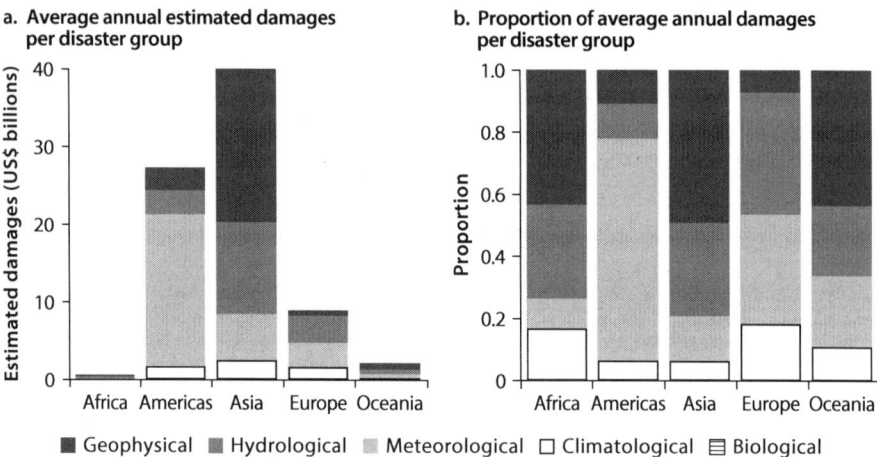

Source: EM-DAT, International Disaster Database, Centre for Research on the Epidemiology of Disasters, Brussels, http://www.emdat.be/natural-disasters-trends.

Co-operation and Development has increased by more than 160 percent, and the risk associated with tropical cyclones has increased by 265 percent. These risks are now growing faster than GDP per capita (Munich Re 2012). As for developing countries, it is well documented that they suffer a 2–15 percent annual loss depending on the country and intensity of the disaster—and as a percentage of GDP, many least developed countries and small island developing states top the list of countries, with damages above 1 percent of GDP (World Bank 2010). On a regional basis, Asia, followed by the Americas, took the biggest hit in estimated damages from 1990 to 2011 (see figure 2.2). Over the past decade, the five countries most frequently hit by disasters were China, India, Indonesia, the Philippines, and the United States.

In light of these huge sums, it is clear that unless we understand and reduce the exposure and vulnerability of people to natural hazards, human development itself is at risk. Emergency responses account for US$63.72 billion—69.9 percent of the annual development spending.[4] Increasing the resilience of the poor could free up hundreds of millions of dollars annually to invest in sustainable development projects.

Warning Systems

The capability of a country to reduce disaster risks depends on three factors:

- Appropriate infrastructure investment
- The ability to provide reliable scientific information on vulnerability, exposure, and predictions of hazards
- The ability to use this information to determine disaster risk and to act to reduce this risk

Developing economies are particularly vulnerable because they are predominantly poor, less resilient, and prone to natural hazards—but also because many lack the capacity to provide risk information to their own citizens and are unable to manage disaster risk effectively. One part of the solution is investing in structural mitigation measures (such as dams and levees), although there are numerous reasons (such as transboundary water issues) such investment is not undertaken in many developing and developed countries (World Bank 2010). Another part is to focus on mapping hazards, to identify vulnerability, to provide accurate risk information, and to build early-warning platforms that enable effective action in response to extreme weather–related events.

Even a few minutes of warning give people time to flee from a flash flood or a tornado or to take refuge from lightning (World Bank 2010). Local authorities use early warnings of tropical cyclones to evacuate large numbers of people to safer locations or to protect them in place in the case of big cities. Long lead times enable people to protect some property and infrastructure. Reservoir operators can reduce water levels gradually to accommodate incoming floodwaters, levees can be inspected and reinforced, and equipment can be positioned for emergency response. Each of these actions depends on access to reliable meteorological and hydrological forecasts and knowledge of the impact of the hazard combined with emergency preparedness.

In every country, NMHSs have the sole or primary mandate to provide the warning services that help protect lives and livelihoods from the adverse effects of meteorological and hydrological phenomena. By international agreement under the auspices of the World Meteorological Organization (WMO) (see box 2.2), NMHSs are expected to provide authoritative warnings of these extreme events, and many NMSs also play a critical role in providing climate information, from monthly and seasonal outlooks to assessments of climate change impacts (Zillman 1999),[5] and in some instances in providing warnings of geophysical hazards.[6] Nationally, NMHSs are the primary operational services providing observations and predictions to support a country's efforts to mitigate climate hazards and the impact of extreme weather events. The most advanced are complex information and communication technology services with access to strong research and development, which support their operational systems (see chapter 3). NMSs also play an important economic role by providing products and services that help different sectors optimize their economic activity in areas such as forecasting demand for electricity and production capacity; forecasting planting and harvesting dates; and managing construction, transportation networks, and inventories.

Many countries have recognized the need to strengthen early-warning systems, but more lag, continuing to underinvest in such systems. Those that have developed effective systems have seen a dramatic reduction in deaths related to weather hazards and have increased their capacity to mitigate hazards. Cuba, for example, has developed an early-warning system that is credited with dramatically reducing deaths for weather-related hazards, such as hurricanes, storm surges, and related flooding (Torres 2012). Based on Meteo France's vigilance system[7]

> **Box 2.2 The World Meteorological Organization**
>
> The World Meteorological Organization (WMO) is an intergovernmental organization of 191 member states and territories, collectively referred to as *WMO Members*.[a] It originated from the International Meteorological Organization, which was founded in 1873. Established in 1950, the WMO is a specialized agency of the United Nations for meteorology (weather and climate), operational hydrology, and related geophysical sciences.
>
> Because weather, the climate, and the water cycle know no national boundaries, international cooperation on a global scale is essential for the development of meteorology and operational hydrology. Such cooperation also allows countries to reap the benefits from the application of the findings. The WMO provides the framework for international cooperation. Under WMO leadership and within the framework of WMO Programs, National Meteorological and Hydrological Services (NMHSs) contribute substantially to protecting life and property against natural disasters, to safeguarding the environment, and to enhancing the economic and social well-being of all sectors of society in areas such as food security, water resources, transport, and health.
>
> The WMO promotes cooperation in (a) establishing networks for making meteorological, climatological, hydrological, and geophysical observations; (b) exchanging, processing, and standardizing related data; and (c) assisting technology transfer, training, and research. It also fosters collaboration between the NMHSs of its Members and furthers the application of meteorology to public weather services, agriculture, aviation, shipping, the environment, water issues, and the mitigation of the impacts of natural hazards.
>
> The WMO facilitates the free and unrestricted exchange of data and information, products, and services in real time or near real time on matters relating to safety and security of society, economic welfare, and the protection of the environment. It contributes to policy formulation in these areas at national and international levels.
>
> a. In this book, states and territories that are members of the WMO are referred to as *WMO Members*. Organizations that belong to the WMO are referred to as *WMO members*.

(Borretti and Degrace 2012), European countries are converging on the development of an alert system, Meteoalarm, which uses common color schemes and symbols to indicate the severity of the weather-related risk in 32 countries.[8] In Asia, the Shanghai Multihazard Early-Warning System showcases a system that enables people to take early action (Tang et al. 2012; see chapter 4).

Forecasting Systems

A reliable forecasting capability is fundamental to an early-warning system for meteorological, hydrological, and climate-related hazards (Zillman 2005a).[9] Accurate forecasting depends on a network of national, regional, and global remote and in situ observations of the atmosphere, oceans, and land operated by NMHSs and their partners (Zillman 2005b). These observations are assimilated by a system of forecast centers, which have differentiated responsibilities for producing global, regional, and national products. This cascading system ensures that large-scale numerical predictions, which require enormous computing power, are created cost-effectively on behalf of all of the Members of WMO. The WMO regional and specialized centers assist the NMHSs in interpreting and downscaling these large-scale products for their own national use. At the national level, the focus is on high fidelity of site- and time-specific information, with an emphasis on delivering forecasts of the impact of the weather and climate and on sharing observational data with global and regional centers.

The foundations of this system were established in the 19th century and have evolved with the advent of satellite observations and skill in numerical weather prediction. Today, most countries take advantage of this communal enterprise through the work of the WMO, which facilitates the global sharing of data, information, and know-how. In large countries, there is often a further cascade within the NMHSs from national centers down through provincial and administrative centers and to municipal and local forecasting offices. At each descending level, data from higher-resolution observation networks and finer-scale forecasts may be used. This arrangement ensures that warnings and other information can be tailored as closely as possible to the users' specific needs.

This enterprise is an excellent example of international cooperation, but it currently depends on the continued investment of advanced countries to sustain the satellites, major computing facilities, and research and development. It also depends on regional investment in adaptation of global products for regional and national application and on national investment in observations and in tailoring of services to the specific needs of a population and economic sectors (Zillman 2005b). Despite the best efforts, this investment in the weather enterprise has not translated into better services in many of the poorest countries. Often, the countries that are most vulnerable to the impact of climate and weather extremes have the least functional NMHSs. The gap between the most advanced NMHSs and those in developing and least developed countries continues to widen.

Arresting the Decline of NMHSs

Over the past 15–20 years, the situation of many NMHSs in developing and least developed countries has worsened, primarily because of underfunding, low visibility, economic reforms, and in some instances military conflict. As a result, many NMHSs do not function well, with some lacking the capacity to provide even a basic level of service. Observation networks have deteriorated, technology is outdated, modern equipment and forecasting methods are lacking, the quality of services is poor, support for research and development is insufficient, and the workforce has been eroded (with many NMHSs lacking trained specialists).[10] As a result, substantial human and financial losses have occurred, which could have been prevented if meteorological and hydrological agencies were more developed. Moreover, there are greater requirements now for NMHSs—for observation, forecasting, and service delivery—to support efforts to make development climate resilient.

A WMO survey indicates that "there are widespread deficiencies in hydrometeorological observing networks, telecommunications, and informatics systems in Africa and very limited … capacities in data management and product customization. [NMHSs] hazard warning capacities are uneven, even nonexistent in some countries, while warning programmes often do not address all significant meteorological and hydrological hazards" (WMO 2008, 21). The survey shows that many NMHSs have limited financial resources to sustain their operation and a weak legal mandate, which may not allow them to generate revenues or provide revenue-generating services.

In Central Asia, for example, observation systems deteriorated dramatically after 1985 (table 2.1). In the Kyrgyz Republic, the number of meteorological stations has been cut by 62 percent, and in Tajikistan, the number of hydrological stations and posts has been cut by 41 percent. In both of these countries and Turkmenistan, upper-air observations—which are very important for forecasting but are expensive—have been completely abandoned.

In addition, WMO indicates that many NHSs in developing countries are in serious decline, with decreasing capacity to monitor river flows and to forecast floods. Difficulties of access, large velocity ranges, inadequate hydraulic conditions, and scarce hydrometric equipment and trained personnel are common throughout the world.

The problems of effective hydrological monitoring are further compounded by the need to consider transboundary river flows in river basins. This need has led to the establishment of regional approaches, such as the Mekong River Commission, which takes responsibility for gauging the main stream of the Mekong on behalf of Cambodia, the Lao People's Democratic Republic, Thailand, and Vietnam and for providing flash flood guidance daily during the rainy season and weekly during the dry season. In turn, each country is responsible for monitoring the river's tributaries within its national boundaries (see box 2.3).

Table 2.1 Deterioration of Hydrometeorological Observation Networks in Central Asia

	Kyrgyz Republic		Tajikistan		Turkmenistan	
Component of observation network	Number, 2008	Percentage of reduction since 1985	Number, 2008	Percentage of reduction since 1985	Number, 2008	Percentage of reduction since 1985
Meteorological stations	32	62	57	22	48	52
Hydrological stations and posts	76	48	81	41	32	45
Upper-air observations	0[a]	100	0[b]	100	0[c]	100
Meteorological radars	0[d]	100	1	75	0[e]	100
Agromet observation stations	31	55	37	46	48	15

Source: World Bank 2009.
a. Before 1985, the Kyrgyz Republic had three operational upper-air stations.
b. Before 1985, Tajikistan had four operational upper-air stations.
c. Before 1985, Turkmenistan had six operational upper-air stations.
d. One radar was in pilot operation in the Kyrgyz Republic before 1985.
e. One radar was in operation in Turkmenistan before 1985.

Box 2.3 Shortfalls in the Ability of the Lao People's Democratic Republic to Monitor River Flows

In the Lao PDR Department of Meteorology and Hydrology, the hydrological forecast office retrieves manual data from field observers and makes these data digitally available to the Mekong River Commission. The commission uses the data to provide flood guidance for the main stream of the Mekong River to all member countries. Just digitizing analog records—including daily records from the observation network as well as historic records—is a full-time job for several staff members.

However, a recent World Bank review team found that the current observation networks are in poor condition. In fact, at the sites visited, the networks are operated entirely manually, with observers reading staff gauges. Automation is limited to the HYCOS[a] stations, which are included in the national part of the commission's basinwide network. Station data are transmitted by telephone or high-frequency radio with the inherent problem of errors introduced in the data transmission process. Most observers are part-time staff members with low salaries and limited incentives to perform their duties. At one of the sites visited, rainfall measurements were not undertaken while some data were sent to headquarters (the data were produced without taking any measurements).

a. HYCOS (Hydrological Cycle Observing System) is a component of the WMO's WHYCOS (World Hydrological Cycle Observing System).

The problem is not limited to NMHSs alone. The WMO regional, specialized, and global centers should play a significant role in helping countries reach a high level of service. However, investment in these centers is limited, and the on-demand human guidance from a regional or specialized center that could be available to a country's NMS or NHS is often not.

Making matters worse is that development funding often takes a short-term focus and centers on national need without considering the benefit of international cooperation. Also, there is often self-interest in much of the donor support, requiring the beneficiary to adopt a particular item of equipment or information technology system that may not fit well with the NMHSs' own strategy or capacity. For example, there are numerous instances of unused or poorly used automatic weather stations or multiple donations of the same type of equipment without the concept of network design or actual need. Many past investments in NMHSs failed to have a significant effect because they focused on providing equipment rather than considering the overall institutional reform and capacity building necessary to provide a modern level of services or because they failed to take into account the additional costs of operating and maintaining new equipment, new infrastructure, or both. Most investments have also focused on country-level assistance only. Although critical, such assistance is just a partial solution and one that may not be sustainable if regional capabilities are weak or no effort is made to access them.

In addition, a vicious circle sometimes develops when NMHSs—especially in developing countries—run into trouble transforming and translating scientific information into something that is actionable by different parts of society. The problem is that without this function, the capacity to influence the decisions of society and the government is severely impaired, which, in turn, contributes to the organizations' declining influence and budget support. A frequent criticism from end users of NMHSs is that they do not understand the information provided and that little care has been given to making NMHSs' products useful. That said, it is also important to find ways to manage unrealistic expectations of user communities and governments for predictive skills that are beyond those that that science can deliver (Zillman 2005b).

Unfortunately, a series of economic policy shifts at the end of the 20th century, in both developed and developing countries, has resulted in public policies that are far less supportive of the concept of professional NMHSs that are exclusively funded by the government (Zillman 2005b). There is a growing sentiment among many governments that the public sector should raise revenue from the sale of services to other government departments and to the public. This expectation presents significant threats to the continued free and unrestricted exchange of information that the WMO has always pursued (Jean et al. 1999; Zillman 1999; see box 2.4).

The encouraging news, however, is that practices that restrict data are increasingly at odds with various international and national directives. For example, European governments are increasingly recognizing the benefits of making their data open and reusable. The wide availability of public sector information can be a key driver in developing content markets in Europe, which can generate new businesses and jobs and can provide consumers with more choice and value for their money.[11] One objective of the European Union's Digital Agenda for Europe is to facilitate the wider deployment and more effective use of digital technologies to enable Europe to address its key challenges and provide Europeans with a better quality of life (for example, by opening up access to content). One element of this objective is INSPIRE (Infrastructure for Spatial Information in the European Community). The INSPIRE directive creates a spatial data infrastructure that allows environmental spatial information to be shared among public sector organizations and better facilitates public access to spatial information across Europe. The regulations of the INSPIRE program encourage data sharing (see box 2.5). Thus, at least in Europe, it is increasingly difficult for NMSs to restrict access to their weather and climate data sets. Hence, the program should engender higher-value-chain service development. For example, in 2007, the Norwegian Meteorological Institute established a free data policy.[12]

Even so, the bottom line is that the leading role of the NMHSs as public service institutions is threatened. NMHSs should have significant roles to play in preventing and reducing meteorological disasters; however, they are often excluded from the civil protection community. Even if warnings are issued, they may carry little authority. Establishing this authority requires government commitment and a legal and regulatory framework that spells out the roles and responsibilities of all stakeholders (see chapter 5). In addition, the absence of a stable funding mechanism for regional and global meteorological and hydrological infrastructure is a significant impediment to national development.

Box 2.4 Threats to the Free and Unrestricted Exchange of Data

The most fundamental principle of World Meteorological Organization (WMO) cooperation is that of free and unrestricted exchange of meteorological, hydrological, and related data and products between members—especially between their National Meteorological and Hydrological Services (NMHSs)—to enable the provision of the best possible weather, climate, and water services at minimal cost in every country. The imperative of international cooperation originated from the need for weather data from distant locations to forewarn of approaching dangerous storms to ships at sea. For almost 130 years, international cooperation in collecting and exchanging meteorological and related information has supported the safety, security, and general welfare of the communications of all nations.

The free and unrestricted exchange of data is now enshrined in Resolution 40 of the 12th World Meteorological Congress and Resolution 25 of the 13th World Meteorological Congress. These resolutions define essential data that should be shared between Members so that each can provide basic forecasts and services (Resolution 40) and flood monitoring (Resolution 25).[a] Because WMO Members are countries and territories, not just NMHSs, these resolutions also apply to specialized observation networks set up by other national agencies for agriculture, energy, and other purposes.

But no legal mandate requires WMO Members to comply with these resolutions, and there is no consistent application of the policy, with many countries actually having more essential and additional data than are shared through the WMO Information System (WIS).[b] This issue is becoming more prominent as countries try to increase their resilience to climate change. Since the requirement for climate information is more extensive than the data covered by Resolutions 40 and 25, countries need to be encouraged to share these additional climate-relevant data so that more accurate regional, national, and local climate predictions, based on large-scale climate simulations, can be determined.[c]

Exacerbating matters, economic policy shifts have resulted in a growing expectation among many governments that the public sector should raise revenue from selling services to other government departments and the public. One unfortunate consequence has been a focus on charging for data, which leads to restricted access. This focus also results in a tendency to strictly interpret Resolutions 40 and 25 as limiting the amount of data shared between WMO Members, which has the unforeseen consequence of limiting the utility of meteorological and hydrological forecasts in those countries where data are most limited. John Zillman, a former president of the WMO, has suggested that these developments have shaken the basic architecture of international meteorology, with many National Meteorological Services (NMSs) being stopped in their tracks on their earlier highly promising road to progress (Zillman 2003).

a. The WMO is reviewing these resolutions to determine their applicability to the Global Framework for Climate Services.
b. The WMO's Global Telecommunication System (GTS) is being upgraded to WIS, which enables access to new data sources in more user-friendly form. WIS replaces the point-to-point data push system of the GTS with a cloudlike network of global information system centers.
c. Principle 6 of the Global Framework for Climate Services states, "The Framework will promote the free and open exchange of climate-relevant observational data while respecting national and international data policies."

Box 2.5 The European INSPIRE Program

In some countries, public data policies are becoming less restrictive. The United States has always had an open data policy under which all publicly funded data are available for reuse without restriction. Some European countries also have such a policy, and through European Union directives, more countries in Europe are making more publicly funded data freely available (Gray 2012).

In Europe, the INSPIRE (Infrastructure for Spatial Information in the European Community) directive entered in force in 2007, establishing an infrastructure for spatial information to support the European Union's environmental policies, along with policies or activities that may affect the environment. The main points of the INSPIRE metadata regulation (2008), as summarized on the INSPIRE website (http://inspire.jrc.ec.europa.eu/index.cfm/pageid/62) are as follows:

> Metadata must include the conditions applying to access and use for Community institutions and bodies; this will facilitate their evaluation of the available specific conditions already at the discovery stage.

> Member States are requested to provide access to spatial data sets and services without delay and at the latest within 20 days after receipt of a written request; mutual agreements may allow an extension of this standard deadline.

> If data or services can be accessed under payment, Community institutions and bodies have the possibility to request Member States to provide information on how charges have been calculated.

> While fully safeguarding the right of Member States to limit sharing when this would compromise the course of justice, public security, national defence or international relations, Member States are encouraged to find the means to still give access to sensitive data under restricted conditions (e.g., providing generalized datasets). Upon request, Member States should give reasons for these limitations to sharing.

Obviously, strengthening the NMHSs in the most vulnerable countries and fully using the WMO to support NMHSs is a "no regrets" strategy. But it must be done on a scale commensurate with the problem. The strategy should be holistic, end to end, and not piecemeal. It should include

- A modernized observation infrastructure
- A forecasting production system that can operate on demand (usually 24 hours a day, 7 days a week)
- Forecasters who can deliver warnings of high-impact events; work with disaster management and other stakeholders; and have expert knowledge of the needs of weather-, climate-, and water-sensitive sectors

A sound risk reduction strategy for national investments includes ensuring the quality of the regional network of WMO centers and using the guidance and know-how that can be channeled through these centers to support NMHSs. This support should include (a) access to observations, (b) numerical model products, (c) on-demand human operational guidance in case of high-impact events, (d) on-the-job training in all aspects of the operations, and (e) institutional "twinning" that pairs more advanced NMHSs with developing NMHSs over a long period. In many developing and least developed countries, this type of overhaul is a major undertaking (World Bank 2009).

What the World Bank Can Do

Before the mid-1990s, the World Bank's approach to strengthening hydrometeorological services involved fragmented efforts to patch up services by supplying individual sensors and partial systems. But since then, the focus has been on a more holistic approach,[13] and today, most efforts involve modernizing entire NMHSs—through strengthening institutions, improving observation networks and forecasting, and enhancing service delivery. In collaboration with the WMO, the World Bank has an advisory role to help convince governments of the great societal and economic significance of weather, climate, and hydrological information and services and of the importance of making meteorological and hydrological agencies the center of this support. It is also helping NMHSs raise their profile in their respective governments by using the results of economic assessments, cost-benefit analyses,[14] and analytical work; identifying priority investment needs; and facilitating financial support.

These modernization activities can be an integral part of larger projects in disaster reduction, water resource management, agricultural support, and public health improvement. Currently, more than a dozen operations are in various stages of implementation, and many more are at the conceptual stage. To help strengthen NMHSs, the World Bank has created a support program—GFDRR Hydromet—which is part of the Global Facility for Disaster Reduction and Recovery (GFDRR).[15] This program, formerly known as Strengthening Weather and Climate Information and Decision-Support Systems, was launched in May 2011 as a vehicle for coordinating many of these activities (see box 2.6).

How much will this modernization cost? In the developing world, more than 100 countries—over half of them in Africa—need to modernize their NMHSs. A conservative estimate of high-priority modernization investment needs in developing countries is between US$1.5 billion and US$2 billion—or perhaps even more. This estimate includes minimum upgrade packages of infrastructure monitoring, capacity building, and institutional strengthening. In addition, a minimum of US$400 million to US$500 million per year is necessary to support operations of the modernized systems (staff costs plus operating and maintenance costs), although national governments are expected to cover those recurrent costs.

But the sums required greatly exceed traditional national budget instruments, often by more than 10 times. For example, table 2.2 shows the estimated costs of modernization of NMHSs in Eastern Europe and Central Asia by how much

Box 2.6 The World Bank's GFDRR Hydromet Program

In 2006, the Global Facility for Disaster Reduction and Recovery (GFDRR) was set up as a partnership of 41 countries and 8 international organizations committed to helping developing countries reduce their vulnerability to natural hazards and adapt to climate change. As part of this program, the World Bank created GFDRR Hydromet in 2011. The program has three pillars:

- *Analytical support and knowledge management.* This pillar involves building analytical knowledge of weather and climate information systems through reviews of weather and climate information and disaster-support (WCIDS) systems; tracing development trends on global, regional, and national levels; assessing knowledge available for sector application; and analyzing how to achieve sustainability of WCIDS systems and increase their societal value. Reviews will draw on input from a network of experts from NMHSs and international, academic, and private sector organizations.
- *Capacity building and technical assistance.* This pillar assists country clients and GFDRR and World Bank teams working on disaster risk management, climate change adaptation, and other objectives in which WCIDS system capacity plays an important supporting role. Workshops, training sessions, and study tours are prepared, and direct consultations and advisory assistance are provided to interested teams. Outputs may include (a) training courses and materials, (b) agreed good practices for investment in WCIDS systems and NMHSs, and (c) identification of innovative practices in using weather and climate information to improve early-warning systems and reduce disaster risk.
- *Portfolio development and operations.* This pillar provides support to GFDRR and World Bank teams that identify major opportunities to advance the disaster risk management agenda by strengthening National Meteorological Services (NMSs). It provides a special project preparation facility for developing early-warning systems and GFDRR Hydromet projects (or major components under broader programs).

Table 2.2 Assessment of Economic Efficacy of Hydromet Modernization, Europe and Central Asia, 2006–08

Country	Estimated cost of modernization program, (2000 US$ millions)	Cost of modernization as a share of annual NMHS budget (%)	Investment efficiency, across seven years	
			Benchmarking assessment (%)	Sector-specific assessment (%)
Albania	4.0	900	438	320–680
Azerbaijan	6.0	350	430	1,440
Armenia	5.3	1,200	210	1,070
Belarus	11.5	400	530	480–550
Georgia	6.0	1,300	260	1,050
Kazakhstan	14.9	350	540	—
Turkmenistan	19.5	170	413	—
Tajikistan	3.5	175	210	310–450
Serbia	4.4	80	880	690
Ukraine	45.3	600	310	410–1,080

Source: World Bank 2008.
Note: — = not available.

these costs exceed NMHSs' annual budgets and the benefits that could be expected in each country. In all cases, the modernization effort is estimated to have a positive effect—such as in Albania, where the investment efficiency across seven years is between 320 percent and 680 percent irrespective of the method of assessment.

The amount of international support for NMHSs is unclear, but it likely is significantly below high-priority needs. We estimate that donors annually allocate roughly US$100–US$200 million to support NMHSs. Furthermore, international support and investment efforts in NMHSs' modernization in developing countries so far have been largely unsuccessful owing to (a) lack of government and development agencies' understanding of the value of the NMHSs and lack of commitment to maintain NMHSs' operations; (b) preoccupation with project time-scale installation of hardware without adequate provision for training, ongoing maintenance, consumables, and other continuing technical support; (c) multiplicity of uncoordinated projects from different donors, each with its own assistance policies, objectives, and equipment suppliers, without sufficient regard to the individual NMHSs' needs, circumstances, and priorities; and (d) technical complexity of the projects.

But as high as the costs of NMHSs' modernization and operation are, they are considerably lower than the expected benefits from prevented losses of life and property and from improved economic performance. In the next chapter, we look at how the global weather, climate, and water enterprise is structured; the special role of the NMHSs; and the critical need to modernize them.

Notes

1. Titled "The Sendai Report: Managing Disaster Risks for a Resilient Future," GFDRR (2012) was prepared to inform the Development Committee at the 2012 World Bank annual meetings and to support discussion at the Sendai Dialogue, a special event supported by the government of Japan and the World Bank.
2. The *2011 Global Assessment Report on Disaster Risk Reduction: Revealing Risk, Redefining Development* (UNISDR 2011a) finds that although global flood mortality risk has decreased since 2000 in the Middle East and North Africa, it is still increasing, thus indicating that growing exposure is outpacing reductions in vulnerability (World Bank 2012).
3. In addition to their meteorological and hydrological responsibilities, many National Meteorological and Hydrological Services (NMHSs) are also responsible for issuing tsunami warnings. These infrequent events have become an integral component of multiple-hazard warning systems.
4. Amount is from the GFDRR Disaster Aid Tracking Database (constant 2009 US$).
5. Zillman's article describes the purpose and structure of the National Meteorological Services (NMS) as a public institution. It provides a comprehensive summary of how NMSs are organized and are related to one another through the World Meteorological Organization (WMO).
6. The Japan Meteorological Agency and the Thai Meteorological Department, for example, also monitor earthquakes and issue tsunami warnings.

7. The vigilance system is described on Meteo France's website at http://france.meteofrance.com/vigilance/Accueil.
8. See the Meteoalarm website at http://www.meteoalarm.eu.
9. Zillman (2005a, 224–25) reiterates the centerpieces of meteorological public policy that have remained unchanged for more than a century, namely, "In order for governments to fulfill their most fundamental obligations to protect the safety and security of their citizens by warning of threatening meteorological events, every country must have in place a basic national meteorological infrastructure of a kind that neither volunteer effort nor market force will provide; and for every country to make best use of its meteorological infrastructure and data in meeting the needs of its own citizens, it must have access to comparable data from other countries as well."
10. The World Bank (2009) summarizes the NMS modernization program for Central Asia, which involves strengthening both national and regional capacity.
11. This area goes well beyond what was envisaged in Resolution 40 by actively encouraging the commercial exploitation of public data.
12. The policy is outlined on the Norwegian Meteorological Institute's website at http://met.no/English/Commercial_Services.
13. The World Bank (2012) describes the investments that have been made in NMSs since 1995 and identifies some of the recurring issues that are being addressed in the current program.
14. For example, benchmarking methods and sector-specific assessments have been used where the overall economic data are poor to demonstrate the cost-effectiveness of modernization programs (Tsirkunov et al. 2006).
15. For more information about the program, visit GFDRR's website at https://www.gfdrr.org/hydrometservices.

References

Borretti, Catherine, and Jean-Noel Degrace. 2012. "The French Vigilance System: Contributing to the Reduction of Disaster Risks in France." In *Institutional Partnerships in Multi-Hazard Early Warning Systems*, edited by Maryam Golnaraghi, 63–93. Heidelberg, Germany: Springer.

DFID (U.K. Department for International Development). 2006. *Reducing the Risks of Disasters: Helping to Achieve Sustainable Poverty Reduction in a Vulnerable World*. London: DFID.

GFDRR (Global Facility for Disaster Reduction and Recovery). 2012. *The Sendai Report: Managing Disaster Risks for a Resilient Future*. World Bank, Washington, DC.

Gray, Mike. 2012. "NMS Data Availability Study for World Bank." Met Office, Exeter, U.K.

Guha-Sapir, Debby, Femke Vos, Regina Below, and Sylvain Ponserre. 2012. *Annual Disaster Statistical Review 2011: The Numbers and Trends*. Centre for Research on the Epidemiology of Disasters, Institute of Health and Society, and Université Catholique de Louvain, Brussels, Belgium.

Jean, Michel, Bruce Angle, David Grimes, and John Falkingham. 1999. "Structure and Evolution of National Meteorological and Hydrological Services: International Comparisons." *WMO Bulletin* 48 (2): 159–65.

Munich Re. 2012. *Topics: Annual Review—Natural Catastrophes, 2011*. Munich, Germany: Munich Re.

Tang, Xu, Lei Feng, Yongjie Zou, and Haizhen Mu. 2012. "The Shanghai Multi-Hazard Warning System: Addressing the Challenge of Disaster Risk Reduction in an Urban Megalopolis." In *Institutional Partnerships in Multi-hazard Early Warning Systems*, edited by Maryam Golnaraghi, 159–79. Heidelberg, Germany: Springer.

Torres, José Maria Rubiera. 2012. "The Tropical Cyclone Early Warning System of Cuba." In *Institutional Partnerships in Multi-Hazard Early Warning Systems*, edited by Maryam Golnaraghi, 9–28. Heidelberg, Germany: Springer.

Tsirkunov, Vladimir, Alexander Korshunov, Marina Smetanina, and Sergey Ulatov. 2006. "Assessment of Economic Efficiency of Hydrometeorological Services in the Countries of the Caucasus Region." World Bank, Moscow.

UNISDR (United Nations International Strategy for Disaster Reduction). 2011a. *2011 Global Assessment Report on Disaster Risk Reduction: Revealing Risk, Redefining Development*. Geneva, Switzerland: United Nations.

———. 2011b. "Preparing for RIO+20: Redefining Sustainable Development." Discussion Paper, UNISDR, Geneva, Switzerland, October 10.

Van Aalst, Maarten. 2006. *Managing Climate Risk: Integrating Adaptation into World Bank Group Operations*. Washington, DC: World Bank.

WMO (World Meteorological Organization). 2008. *Capacity Assessment of National Meteorological and Hydrological Services in Support of Disaster Risk Reduction: Analysis of the 2006 WMO Disaster Risk Reduction Country-Level Survey*. Geneva, Switzerland: WMO.

World Bank. 2008. "Weather and Climate Services in Europe and Central Asia: A Regional Review." Working Paper 151, World Bank, Washington, DC.

———. 2009. *Improving Weather, Climate, and Hydrological Services Delivery in Central Asia (Kyrgyz Republic, Republic of Tajikistan, and Turkmenistan)*. Russia Country Office, World Bank, Washington, DC.

———. 2010. *Natural Hazards, Unnatural Disasters: Effective Prevention through an Economic Lens*. Washington, DC: World Bank.

———. 2012. *Strengthening Weather and Climate Information and Decision Support Systems (WCIDS): The World Bank Portfolio 1995–2011*. Washington, DC: World Bank.

Zillman, John W. 1999. "The National Meteorological Service." *WMO Bulletin* 48 (2): 129–59.

———. 2003. "The State of National Meteorological Services around the World." *WMO Bulletin* 52 (4): 360–65.

———. 2005a. "The Challenges for Meteorology in the 21st Century." *WMO Bulletin* 54 (4): 224–29.

———. 2005b. "Real Time Data Requirements of National Meteorological Services (NMHSs) and Their Users." Paper presented at the National Oceanic and Atmospheric Administration's National Weather Service International Session on Addressing Data Acquisition Challenges, San Diego, CA, January 6–7.

CHAPTER 3

National Meteorological and Hydrological Services

In This Chapter
The World Meteorological Organization (WMO) and National Meteorological and Hydrological Services (NMHSs) are, respectively, the international and national entities that underpin the global meteorological and hydrological forecasting and warning systems. Governments need to support their NMHSs to (a) maintain and operate national meteorological and hydrological networks, (b) provide forecasts of routine and extreme meteorological and hydrological events, (c) issue timely and accurate early warnings, and (d) communicate critical information to stakeholders and the public.

Introduction
The global meteorological and hydrological enterprise consists of all entities contributing to production, delivery, and use of weather, climate, and water[1] information and services. It includes government, academia, and the private sector; intergovernmental organizations; and specialized global and regional meteorological and hydrological forecasting, training, and support centers.

At the heart of this system are the National Meteorological Services (NMSs) and National Hydrological Services (NHSs)—collectively referred to as NMHSs (Zillman 1999). They are responsible for providing government and civil society with information to protect lives and property and for improving economic well-being through timely forecasts and warnings of high-impact meteorological and hydrological events (figure 3.1).

Academia focuses on better understanding the physical and social sciences: (a) developing new observation and forecasting techniques; (b) quantifying the impacts of weather, climate, and water on society; (c) communicating risk to decision makers; and (d) educating professionals in both the provider and user communities to be able to work with each other.

Figure 3.1 Activities of a Typical National Meteorological Service

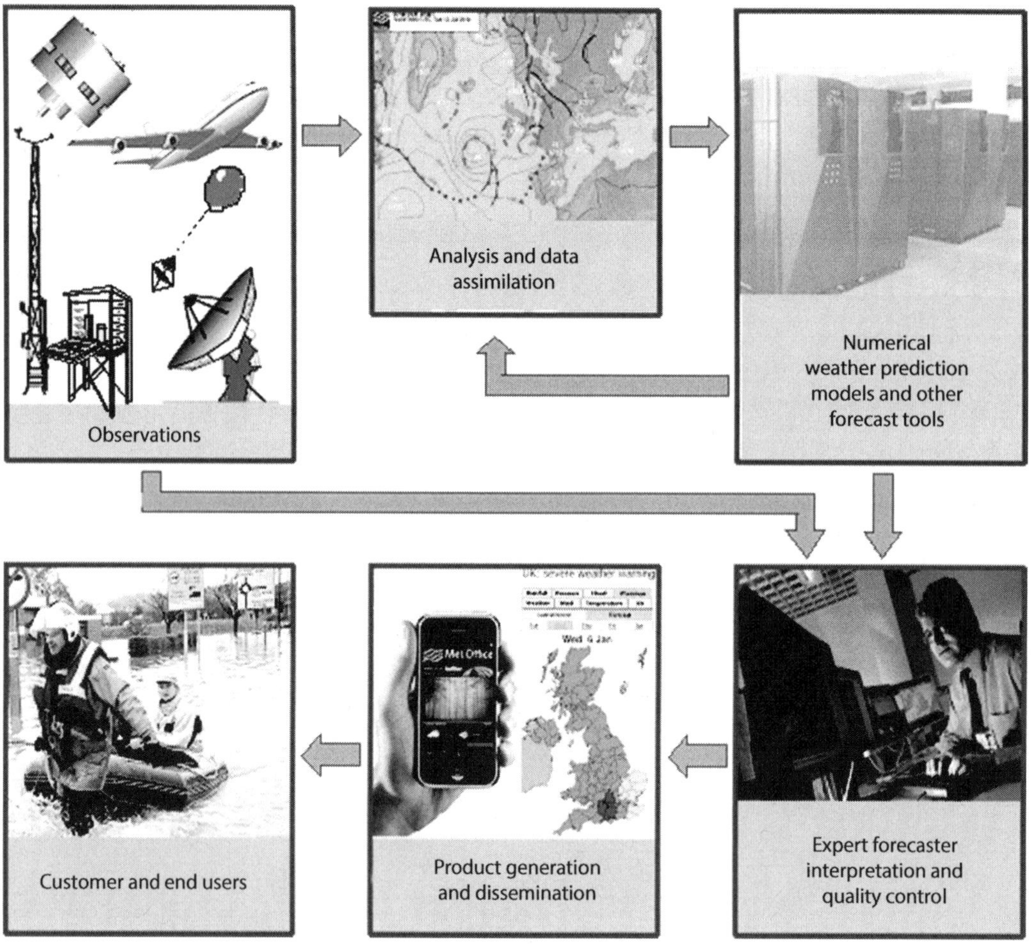

Source: Gray 2012.

In addition, private sector commercial weather, climate, and water services exist in many parts of the world. They often focus on specific users and sometimes deliver public forecasts through print, broadcast, mobile, and other media. Tensions between the public and private sectors can arise if roles and responsibilities have no distinct separation, but their roles are typically complementary, which strengthens the overall global enterprise. This chapter explores this vast enterprise before focusing on the NMHSs and the enormous challenges they face in modernizing.

The Global Weather, Climate, and Water Enterprise

Because weather, climate, and the water cycle know no national boundaries, international cooperation is essential for developing meteorology and operational hydrology, as well as for reaping the benefits from their application. The World

> **Box 3.1 The World Meteorological Organization's Structure**
>
> The World Meteorological Congress, the supreme body of the World Meteorological Organization (WMO), brings together the delegates of its Members, led by the permanent representative (PR), once every four years to determine general policies for fulfilling the organization's purposes. Under the World Meteorological Convention, the PR of a Member of the WMO is usually the head of a National Meteorological Service or National Hydrological Service.
>
> The Executive Council, the executive body of the organization, meets annually and is responsible to the World Meteorological Congress for coordinating the organization's purpose and using its budgetary resources. It is composed of 37 directors of National Meteorological or National Hydrological Service (NHSs) serving in their personal capacities on behalf of the entire organization.
>
> There are six regional associations: Region I (Africa); Region II (Asia); Region III (South America); Region IV (North America, Central America, and the Caribbean); Region V (South-West Pacific); and Region VI (Europe).
>
> There are eight technical commissions: (a) Basic Systems, (b) Instruments and Methods of Observation, (c) Atmospheric Sciences, (d) Aeronautical Meteorology, (e) Agricultural Meteorology, (f) Oceanography and Marine Meteorology, (g) Hydrology, and (h) Climatology.

Meteorological Organization—which was created in 1950 and became a specialized agency of the United Nations system in 1951 (headquartered in Geneva, Switzerland)—provides the framework for such cooperation (see box 3.1).

The purpose of the WMO is sixfold:

- To facilitate worldwide cooperation in establishing the networks of stations for making of meteorological observations
- To promote the establishment and maintenance of systems for the rapid exchange of meteorological information
- To promote standardization of meteorological and related observations and to ensure the uniform publication of observations and statistics
- To further the application of meteorology to aviation, shipping, water problems, and other human activities
- To promote activities in operational hydrology and to further close cooperation between meteorological and hydrological services
- To encourage research and training in meteorology

How does the WMO manage the analysis and forecasting of the state of the atmosphere? The process, coordinated by the World Weather Watch, combines three distinct activities: observing systems, telecommunication systems, and global data processing and forecasting systems (figure 3.2).[2]

Figure 3.2 World Meteorological Organization Global Network

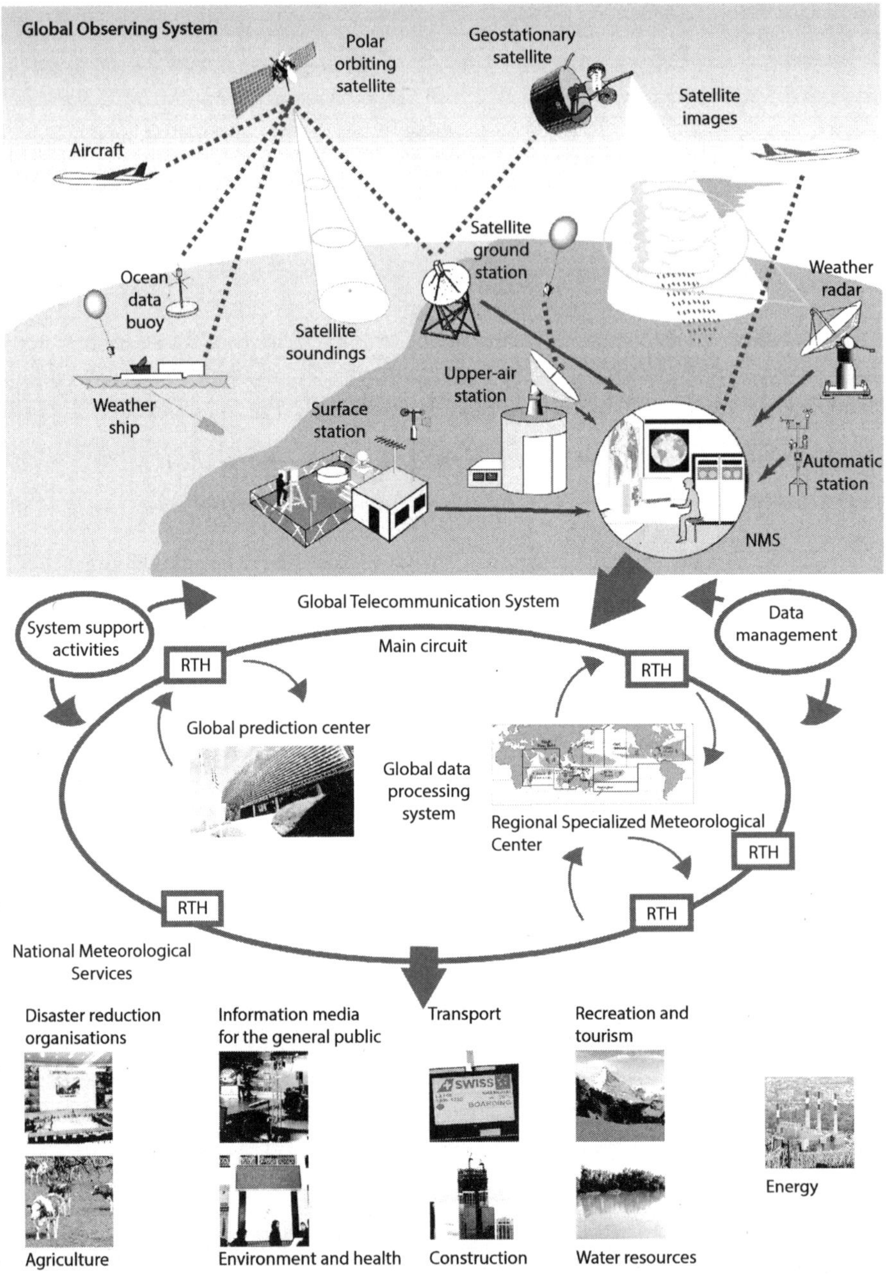

Source: World Meteorological Organization.
Note: NMS = National Meteorological Service; RTH = Regional Telecommunication Hub. The network comprises the Global Observing System, Global Telecommunication System, and Global Data Processing and Forecasting System—made up of 3 World Meteorological Centers and 40 Regional Specialized Meteorological Centers and National Meteorological and Hydrological Services.

Observing Systems

Given that observations are needed of the entire atmosphere, the oceans, and the land surface, the World Weather Watch uses the following equipment:

- Satellite remote sensing for high-resolution spatial information
- Upper-air observations from balloon-borne sensors and commercial aircraft for accurate point-source measures of the vertical structure of the troposphere
- Radar for high-resolution and immediate information on precipitation rates and for identification of intense, short-lived weather hazards
- Land and ocean surface measurements for detailed information on the basic state near the surface: temperature, humidity, wind, pressure, and shortwave and longwave radiation (see photo 3.1)

In addition to their direct application, these observations are used to initialize and constrain numerical weather, climate, and water prediction models. In this regard, standardization of meteorological and hydrological observations is very

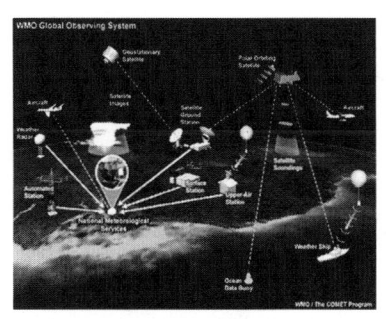

a. Components of the Global Observing System

b. Ground-based temperature measurement, Vietnam

c. S-band Doppler radar, Cambodia

d. Water-level measurements, Nepal

Photo 3.1 Multiple observation systems.

Source: Panel a photo is courtesy of the World Meteorological Organization Integrated Global Observing System.

important and is overseen by the WMO Commission for Instruments and Methods and Observation (CIMO) and the WMO Instruments and Methods of Observation Program. These bodies set technical standards, quality-control procedures, and guidance for using meteorological instruments and observation methods to promote worldwide standardization.[3] This oversight also permits unbiased assessment of new technologies before widespread introduction.

The WMO Integrated Global Observing System (WIGOS)[4] is fostering the evolution of one comprehensive coordinated system from three separate observation systems: (a) the Global Observing System (GOS),[5] which is the basic meteorological observation system; (b) the Global Atmospheric Watch (GAW),[6] which measures the chemistry of the atmosphere; and (c) the World Hydrological Cycle Observing System (WHYCOS).[7] WIGOS is also expected to integrate observations external to NMHSs into the system and to improve the exchange of data with common formats and metadata. WIGOS is essential to enable WMO Members, in collaboration with national agencies, to meet countries' observational requirements to improve timely advisories and early warnings of extreme weather and climate events. It also underpins the improvement in weather, climate, water, and environmental monitoring and forecast services and in adaption to and mitigation of climate change, especially in developing countries and the least developed countries (WMO 2011).

In addition, systematic climate observations are needed to address the challenges of climate change. For this reason, the Global Climate Observing System[8] (GCOS) was established in 1992, in parallel with the adoption of the United Nations Framework Convention on Climate Change (UNFCCC). Its purpose is to ensure that climate-relevant measurements of the atmosphere, ocean, and land are made systematically and continually and are disseminated to all interested users. The aims and requirements of systematic observation are specified both in the UNFCCC and in the subsequent Kyoto Protocol. At the national level, the GCOS Implementation Plan (WMO 2010) requires countries to coordinate climate-relevant measurements (see box 3.2).

Telecommunication Facilities

Data from this "system of systems" is shared through the WMO Information System (WIS), which is the single coordinated global infrastructure responsible for telecommunication and data management. WIS is designed to extend the existing Global Telecommunication System and WMO Members' ability to collect and disseminate data and products. Owned and operated by the WMO Members, WIS is the core information system used by the WMO community, providing links for WMO and supported programs associated with weather, climate, water, and related products and services.

Data Processing and Forecasting Centers

The Global Data-Processing and Forecasting System (GDPFS)[9] is the means by which global numerical weather prediction (NWP) is made possible (see box 3.3 technical insight). Its main purpose is to prepare and make available

Box 3.2 How Switzerland Supports the Global Climate Observing System

Following the Swiss Parliament's ratification of the Kyoto Protocol in 2003, national coordination of the activities of the Global Climate Observing System (GCOS) was intensified, leading to the establishment of the Swiss GCOS Office at the Federal Office of Meteorology and Climatology (MeteoSwiss) in 2006.[a] Today, the Swiss GCOS Office is responsible for coordinating all climate-relevant measurements in Switzerland made by various institutions. In close collaboration with universities, federal offices, private institutions, and research institutes—and advised by a steering committee of policy makers and scientists—the Swiss GCOS Office has recently compiled an inventory of key climate measurement time series in Switzerland (Seiz and Foppa 2007).

Thanks to this agencywide cooperation, analyzing potential discontinuities in observation networks and assisting partners in the long-term planning of systematic climate observation are possible. To sustain and foster the dialogue, a GCOS roundtable is organized once a year to discuss ongoing projects and activities. In doing so, the Swiss GCOS Office facilitates the information exchange between various institutions in Switzerland and acts as an interface between providers and users of climate-related data.

Given the need to design national adaptation strategies in response to climate variability and change, the experience gained from these coordination mechanisms will help in implementing national service delivery strategies for climate services as part of the Global Framework for Climate Services.[b]

a. See the Swiss GCOS website at http://www.gcos.ch.
b. See the WMO website at http://www.wmo.int/pages/gfcs/index_en.php.

---

meteorological analyses and forecasting products in the most cost-effective way. Global NWPs products are generated at a few specialized centers, which have the computing power and technical staff to run these models. In many centers, these models now run at such high spatial resolution (better than 15 kilometers horizontal resolution) that they can be used directly by NMHSs in their own forecast production systems. Alternatively, they are used to provide inputs for limited area models, which assimilate local data that are unavailable to the global system, to provide more precisely tailored forecasts usually on short time-scales. The GDPFS operates through the world meteorological centers (WMCs); see box 3.4 technical insight), regional and specialized meteorological centers[10] (see box 3.5 technical insight), and national meteorological centers (NMCs).

The Special Role of NMHSs

At the national level, the key players are the NMHSs, which have a fourfold mission (Zillman 2003a):

- Observing and monitoring the state of the atmosphere, ocean, and inland waters
- Conducting research aimed at understanding, modeling, and predicting the behavior of atmospheric and related processes and phenomena

---

Box 3.3 Technical Insight: Functions of the Global Data Processing and Forecasting System

The real-time functions of the Global Data-Processing and Forecasting System (GDPFS) include

- Preprocessing of data (such as retrieval, quality control, decoding, and sorting of data stored in a database for use in preparing output products)
- Preparing analyses of the three-dimensional structure of the atmosphere with up-to-global coverage
- Preparing forecast products (fields of basic and derived atmospheric parameters) with up-to-global coverage
- Preparing ensemble prediction system products
- Preparing specialized products (such as limited-area very fine mesh short-, medium-, extended-, and long-range forecasts and tailored products for marine, aviation, environmental quality monitoring, and other purposes)
- Monitoring observational data quality
- Postprocessing numerical weather prediction data using workstation and personal computer–based systems to produce tailored value-added products and to generate weather and climate forecasts directly from model output

The non-real-time functions of the GDPFS include

- Preparing special products for climate-related diagnosis (that is, 10-day or 30-day means, summaries, frequencies, and anomalies) on a global or regional scale
- Comparing analysis and forecast products; monitoring observational data quality; and verifying the accuracy of prepared forecast fields, diagnostic studies, and numerical weather prediction model development
- Storing long-term Global Observing System (GOS) data and GDPFS products, as well as verifying results for operational and research use
- Maintaining a continuously updated catalogue of data and products stored in the system
- Exchanging ad hoc information through distributed databases among GDPFS centers
- Conducting workshops and seminars on the preparation and use of GDPFS output products.

- Providing information, forecast, warning, and advisory services to the community at large and to specific user groups
- Fulfilling obligations for international cooperation in data exchange and service provision, under the WMO convention.

By providing a variety of services related to essential weather, climate, and water services,[11] NMHSs contribute to numerous national goals (in order of importance based on a global WMO survey[12]): (a) safety of life and property; (b) reduction of the impact of natural disasters; (c) economic development and prosperity of primary, secondary, and tertiary industry; (d) community health,

Box 3.4 Technical Insight: World Meteorological Organization Global Centers

The three World Meteorological Centers (WMCs)[a]—Melbourne, Moscow, and Washington, DC—are capable of applying sophisticated global numerical weather prediction models (including ensemble prediction systems, or EPS) and preparing the following products for distribution to Members and other Global Data Processing and Forecasting System centers:

- Global (hemispheric analysis) products
- Short-, medium-, extended-, and long-range forecasts and products that have global coverage but are presented separately, if required, for (a) the tropical belt or (b) the middle and high latitudes or any other geographic area
- Climate-related diagnostic products, particularly for tropical regions

The WMCs also (a) carry out verification and intercomparison of products, (b) support the inclusion of research results into operational models and their supporting systems, and (c) provide training courses on the use of WMC products (see WMO 1992).

For climate, there is also a network of global producing centers[b] for long-range forecasts (monthly and seasonal), which provide as a minimum requirement the following products:

- Predictions for averages, accumulations, or frequencies over one-month periods or longer[c]
- Lead time of between zero and four months
- Issue frequency, which is monthly or at least quarterly
- Delivery of graphical images on the global producing center's website or digital data for download
- Variables such as two-meter temperature, precipitation, sea-surface temperature, mean sea-level pressure, 500 hectopascal height, and 850 hectopascal temperature
- Long-term forecast skill assessments

a. To learn more about the WMCs, visit the WMO website at http://ww.wmo.int/pages/prog/www/DPS/gdps-2.html#WMCs.
b. The officially designated World Meteorological Organization (WMO) global producing centers for long-range forecasts are the Bureau of Meteorology in Australia, the Beijing Climate Center of the China Meteorological Administration, the Climate Prediction Center of the U.S. National Oceanic and Atmospheric Administration, the European Centre for Medium-Range Weather Forecasts, the Tokyo Climate Center of the Japan Meteorological Agency, the Korean Meteorological Administration, Météo France, the U.K. Met Office, the Meteorological Service of Canada, the South African Weather Service, the Hydrometeorological Center of the Russian Federation, and the Center for Weather Forecasts and Climate Studies of the National Institute for Space Research in Brazil. Other major centers providing global seasonal forecasts include the International Research Institute for Climate and Society and the Asia-Pacific Economic Cooperation Climate Center in the Republic of Korea. There are also two lead centers: (a) the WMO Lead Center for Long-Range Forecast Multi-model Ensemble, which is jointly coordinated by the Korean Meteorological Administration and the U.S. National Weather Service's National Centers for Environmental Prediction, and (b) the WMO Lead Center for Standard Verification System of Long-Range Forecasts, which is jointly coordinated by the Bureau of Meteorology in Australia and the Meteorological Service of Canada.
c. Typically, anomalies in three-month-averaged quantities are the standard format for seasonal forecasts, and forecasts are usually expressed probabilistically.

recreation, and quality of life; (e) national and international security; (f) preservation and enhancement of the quality of the environment (ranked above national security in 2011); (g) fulfillment of international requirements and commitments; (h) advancement of knowledge and understanding of the natural systems of the planet; (i) efficient planning, management, and operation of government

Box 3.5 Technical Insight: World Meteorological Organization Regional Specialized Centers

The regional specialized centers of the World Meteorological Organization (WMO)[a] are centers with either geographic specialization or activity specialization. They include (a) Regional Specialized Meteorological Centers (RSMCs), (b) Tropical Cyclone Warning Centers (TCWCs), (c) Regional Support Centers, (d) Regional Training Centers (RTCs), and (e) Regional Climate Centers (RCCs). Each of these centers is responsible for distributing information advisories and warnings for the specific program that they are a part of, agreed by consensus of the WMO World Weather Watch.

RSMCs play a vital role in a cascading forecasting process. Numerical weather prediction (NWP) centers (a) provide available NWP and ensemble prediction system products to the RSMCs, which interpret the information received; (b) prepare daily guidance products for the National Meteorological Services' (NMSs) National Meteorological Centers (NMCs); (c) run limited area models to refine products on smaller scales; (d) maintain information on websites; and (e) coordinate with participating NMCs. The RSMCs have differentiated responsibilities for atmospheric radiation, atmospheric sand and dust forecasts, medium-range weather forecasting, tropical cyclone forecasting, and so forth. For example, six RSMCs are dedicated to tropical cyclone prediction, in addition to the TCWCs.[b] Other functions include the following:

- RSMC New Delhi is responsible for issuing tropical weather outlook and tropical cyclone advisories in the region bordering the Bay of Bengal and the Arabian Sea, including Bangladesh, Maldives, Oman, Pakistan, Republic of the Union of Myanmar, Sri Lanka, and Thailand.
- RSMC Beijing is responsible for environmental emergency response within the WMO's Region II. It is required to provide advice, in the form of a basic set of products, on the atmospheric transport of pollutants resulting from nuclear disasters, forest fires, and chemical incidents.
- RSMC Exeter carries out similar functions for Regions I and VI through the operations center of the U.K. Met Office, which incorporates a national environmental emergency monitoring and response center staffed by forecasters with specialized training in nuclear and chemical incidents, airborne animal diseases, smoke from large fires, and volcanic ash.
- RSMC Obninsk carries out the same functions within the Russian Federal Service for Hydrometeorology and Environmental Monitoring (Roshydromet).

Other specialized centers include the European Centre for Medium-Range Weather Forecasts, the African Centre of Meteorological Applications for Development, and the National Centre for Medium-Range Weather Forecasting-New Delhi.

RTCs undertake training in meteorology, hydrology, and related sciences, which are created specifically to meet the requirements of at least half the members of a Regional Association that cannot be met by existing facilities. But a major limitation is that they focus mostly on technical training, not training on delivering services, and they rarely, if ever, include users of meteorological services.

RCCs are intended to be centers of excellence that create regional climate products, including long-range forecasts to support regional and national climate activities. As such,

box continues next page

> **Box 3.5 Technical Insight: World Meteorological Organization Regional Specialized Centers**
> *(continued)*
>
> they are important in strengthening the capacity of WMO members through their NMSs in a given region to deliver better climate services to national users. In general, they should be part of any modernization program that involves strengthening national climate services.
>
> a. The WMO's web page for the centers is http://www.wmo.int/pages/prog/www/DPS/gdps-2.html#WMCs.
> b. The six RSMCs are as follows: (a) for the Southwest Pacific Ocean, RSMC Nadi-Tropical Cyclone Centre, Fiji Meteorological Service (Nadi, Fiji); (b) for the Southwest Indian Ocean, RSMC La Réunion-Tropical Cyclone Centre, Météo France (Réunion Island, French Overseas Department); (c) for the Bay of Bengal and Arabian Sea, RSMC Tropical Cyclones New Delhi, India Meteorological Department (New Delhi, India); (d) for the Western North Pacific Ocean and South China Sea, RSMC Tokyo, Japan Meteorological Agency (Tokyo, Japan); (e) for the Central North Pacific Ocean: RSMC Honolulu Central Pacific Hurricane Center (Honolulu, Hawaii); (f) for the Northeast Pacific Ocean, Gulf of Mexico, Caribbean Sea, and North Atlantic Ocean, RSMC Miami, National Hurricane Center (Miami, Florida). The TCWCs are as follows: (a) for Indonesia, the TCWC-Jakarta, Meteorological and Geophysical Agency; (b) for the Tasman Sea, the TCWC-Wellington, Meteorological Service of New Zealand; (c) for the Solomon Sea and Gulf of Papua, the TCWC–Port Moresby, National Weather Service, Papua New Guinea; (d) for the Coral Sea, the TCWC-Brisbane, Bureau of Meteorology, Australia; (e) for the Arafura Sea and Gulf of Carpentaria, TCWC-Darwin, Bureau of Meteorology, Australia; and (f) for the Southeast Indian Ocean, TCWC-Perth, Bureau of Meteorology, Australia.

and community affairs; (j) provision for the information needs of future generations; and (k) policy setting.[13]

NMHSs make a significant contribution to safety, security, and economic well-being. However, this contribution is rarely quantified, which often results in undervaluation of the importance of the NMS or NHS in supporting a country's capacity to cope with weather- and water-related hazards and in providing the economic benefit of accurate weather, climate, and water information to increase productivity and avoid losses. Hallegatte (2012) estimates that upgrading all hydrometeorological information production and early-warning capacity in developing countries would prevent between US$300 million and US$2 billion in losses owing to disasters, would save an average of 23,000 lives, and would provide between US$3 billion and US$30 billion in additional economic benefits from disaster reduction.

It is essential that one organization in a country, the NMS, be responsible for issuing authoritative weather forecasts and warnings. In some countries, this responsibility extends to other warnings as a part of a multihazard approach, and it may require a more collective governmental response involving several departments or agencies. A single authoritative voice for a particular warning provides consistency and avoids confusion in anticipation of, and during, any disaster. That is not to say that NMSs take on the responsibility for issuing evacuation orders; rather they are the source of authoritative weather and water warning information that forms part of a decision process that involves (a) identifying meteorological and hydrological hazards, (b) alerting government decision makers, (c) communicating early warning to the public and at-risk sectors, and (d) responding in a timely and effective manner. Issuing warnings at short notice of intense atmospheric phenomena, such as tornadoes, should always be the direct responsibility of the NMS, which must have the capacity to communicate directly with the public as well as officials (see chapter 4).

A Snapshot of Hydrological Services

As with meteorological services, the largest component of the hydrological services' mission is to meet its public obligations associated with managing water resources and reducing flood- and drought-related hazards. Indeed, the fundamental influences on an NHS are government policies and national development goals, as well as the information that may be needed to support them. Besides the public, clients are likely to include national disaster management, land-use planning, infrastructure, municipal wastewater treatment, environment, and agriculture—some of which likely operate hydrological departments.

What are the major challenges facing NHSs? Studies by the WMO (2006, 2009a) identified the following factors, which remain current today:

- Sustainable management or stewardship of natural resources and the environment in conjunction with efforts to improve the living conditions of the poor
- Integrated water resource management for sustainability
- Disaster mitigation in cooperation with NMSs in the face of the increasing impact of natural disasters, especially floods and droughts
- Rapid developments in technology that enable NHSs to improve products and services but at significantly increased investment costs and staff retraining
- Expectation that public services should be more accountable, resulting in public sector reform to increase efficiency, effectiveness, and value for money with greater competition for resources in the public sector
- Greater emphasis on the need to manage transboundary river basins
- Social trends, including an increasing role of women in public service

Traditionally, operational hydrology has had a strong focus on surface-water quality and river basin hydrology, but it also includes the quantity and quality of groundwater. In addition, many countries have needed to monitor extreme events, particularly floods, in real time for warning purposes. Today, hydrologists take a much broader view of hydrology, extending it to ecological, biological, and human use of the aquatic system—which means that the activities of many NHSs are increasing. Nevertheless, hydrometric measurement remains a core function.[14]

What exactly is hydrometry? It primarily involves measuring river flows at gauging stations, along with rainfall and groundwater, to provide the foundation for better river and water management strategies (see box 3.6 technical insight). River flows are the combined result of the many climatological and geographic factors that interact within a drainage basin. This integration implies that river flows are the most directly appropriate component in the hydrological cycle for a wide range of applications:

- Assessing and managing water resources (irrigation provision)
- Designing water-related structures (reservoirs, bridges, flood banks, urban drainage schemes, and sewage treatment work)
- Creating flood warning and alleviation schemes

Box 3.6 Technical Insight: Key Types of River Flow Measurements

Indirect Flow Measurement

River flows are normally measured indirectly—relying on the conversion of a record of water level (or stage) to flow using a stage-discharge relation, often referred to as the *rating* or *calibration*. At primary gauging stations, stage is generally measured and recorded against time by instruments actuated by a float in a stilling well. Solid-state loggers are normally deployed to record water level, often with a chart recorder for backup. Where possible, provision is made for the routine transmission of river levels directly to a processing center, most commonly by telephone or satellite communication link.

Stage-Discharge Relation

The stage-discharge relation is obtained either by installing a gauging structure—usually a weir or flume with known hydraulic characteristics—or by measuring the stream velocity using propeller-type current meters or other methods. Acoustic Doppler Current Profilers are increasingly being used, which offer significant advantages (for example, speed of flow assessment and greater safety for the operators) over traditional current meters.

The mean velocity is combined with the cross-sectional area of the river to provide a measurement of flow at points throughout the flow range at a site characterized by its ability to maintain a reasonably stable relationship. The stage-discharge relationship may be disturbed by changes to the hydraulic characteristics of the gauging reach, for example, owing to changes in the bed profile following a flood or the seasonal effect of aquatic plant growth.

Ultrasonic Flow Measurement

For ultrasonic gauging stations, a stable relationship between river level and flow is unnecessary. Flows are computed on site, where the times are measured for acoustic pulses to traverse a river section along an oblique path in both directions. The mean river velocity is related to the difference in the two timings, and the flow is then assessed using the river's cross-sectional area.

Accurate computed flows can be expected for stable river sections and within a range in stage that permits good estimates of mean channel velocity to be derived from a velocity traverse set at a series of fixed depths. Accuracy can be compromised by high concentrations of suspended sediment or heavy weed growth, which can impede the acoustic signal, or by thermal stratification in the water column, which can deflect the acoustic beams.

Electromagnetic Flow Measurement

Flow data from electromagnetic gauging stations may also be computed on site. The technique requires measuring the electromotive force induced in flowing water as it cuts a vertical magnetic field generated by means of a large coil buried beneath the riverbed or constructed above it. This electromotive force is sensed by electrodes at each side of the river and is directly proportional to the average velocity in the cross-section. This method is not common owing to technical, maintenance, and health and safety issues.

Source: Data from Centre for Ecology and Hydrology, http://www.ceh.ac.uk.

- Assessing and developing hydropower potential
- Enhancing both the ecological health of watercourses and wetlands and their amenity and recreational value
- Underpinning water policy initiatives, protocols, and legislation (for national and international application), such as for river basins and integrated water resource management
- Serving a fundamental scientific and educational role in understanding hydrological processes and contributing to a greater public understanding of water and water management issues.

As with many environmental data sets, river flow data assume a particular importance at a time of actual or anticipated change. With climate change expected to have very uneven effects on river flow patterns (in both spatial and temporal terms), observational records are key for identifying, quantifying, and interpreting hydrological trends—a necessary precursor for developing more effective coping mechanisms in future flood and drought episodes.

How accurate are these hydrometric measurements? International standards are followed as much as possible in the design, installation, and operation of gauging stations. Most of these standards include a section devoted to accuracy, and many include recommendations for reducing uncertainties in discharge measurements and for estimating the extent of the uncertainties, which do arise (see box 3.7 technical insight).

National hydrological monitoring programs should aim to provide the authoritative voice on hydrological conditions, placing this information in a historical context and, over time, identifying and interpreting any hydrological trends. Such information is essential for improving water management strategies and for increasing public understanding of hydrological and water resource issues. This information should be disseminated through regular monthly and annual bulletins, occasional reports on floods and droughts, and engagement with the media.

For most countries, such monitoring is a shared responsibility among several measuring authorities. In the United Kingdom, for example, hydrological monitoring is carried out by the Environment Agency, the Scottish Environmental Protection Agency, the Rivers Agency, the British Geological Survey, and the Centre for Ecology and Hydrology (see box 3.8). In the Republic of Yemen, monitoring is the responsibility of the National Water Resources Authority, which works closely with Development Authorities and the Ministry of Agriculture and Irrigation—each of which monitors river flows and has some responsibility for flood and drought warnings. Monitoring may include rainfall, river flows, borehole levels, and reservoir stocks. In contrast, a number of countries combine their meteorological and hydrological services. One example is the National Hydro-Meteorological Service of Vietnam, which has extensive meteorological and hydrological observation networks that are at present being refurbished and extended (see photo 3.2).

Box 3.7 Technical Insight: Ensuring the Usefulness of River Flow Data

In shallow rivers, the accuracy of river flows depends primarily on the precision of the stage measurement. Modern sensing and recording instrumentation is capable of measuring water levels to tolerances of a few centimeters. However, imprecise surveying of the gauging station data or the growth of algae on the crest of a weir can introduce systematic errors. Similarly inadequate maintenance can render even the most up-to-date systems useless (see photo TI3.7.1).

Uncertainties in the stage-discharge relationship can also be substantial, particularly in the extreme flow ranges where relatively few gauges may be available to define the rating. In gauging sections where the channel's hydraulic characteristics are subject to change (for example, owing to erosion or accretion following a flood or to seasonal variations in the growth of aquatic plants), changes in the stage-discharge relation require careful monitoring to minimize consequential uncertainties in computed flows.

The utility or fitness for purpose of river flow data reflects their accuracy but is also strongly influenced by a number of other factors. Even a small proportion of missing data can greatly reduce the ability to derive meaningful summary statistics (for example, annual runoff totals or seven-day minimal).

The nature of hydrometric measurement determines that missing data tend to cluster disproportionately in the extreme flow ranges, reflecting the difficulty of precisely capturing the highest and lowest flows. However, the exceptional importance of extreme flows underlines the need for effective procedures to derive estimates of missing flows. Judgment needs to be exercised in applying such procedures to avoid archiving misleading flow estimates. But in most circumstances, the inclusion of an auditable and suitably flagged estimate—rather than leaving a gap in the record—will produce significant benefits in relation to the overall value of the river flow series.

Photo TI3.7.1 Vegetation obstructing water-level measurements.

Box 3.8 Hydrological Monitoring in the United Kingdom

The United Kingdom's National Hydrological Monitoring Program relies on the active cooperation of many agencies. It uses a subset of the National River Flow Archive gauging stations, National Groundwater Level Archive index wells, and measuring authority and water company reservoirs for analysis within its monthly hydrological summaries. The regional divisions of the Environment Agency, the Environment Agency Wales, the Scottish Environment Protection Agency, the Rivers Agency, and the Northern Ireland Environment Agency provide data on river flow and groundwater levels. The water service companies, the Environment Agency, Scottish Water, and Northern Ireland Water provide details on reservoir stocks.

The Met Office's National Climate Information Centre (NCIC) provides most of the rainfall data. To allow better spatial differentiation, most regional rainfall figures for the United Kingdom are presented for the regional divisions of the precursor organizations of the Environment Agency and the Scottish Environmental Protection Agency. For historical comparisons of national and regional rainfall totals, the NCIC monthly series is normally used. But for England and Wales, the homogenized England and Wales rainfall series developed by the Climatic Research Unit at the University of East Anglia and currently updated by the Met Office's Hadley Centre may be used.

Photo 3.2 River gauging housing located on a tributary of the Mekong River, near Ho Chi Minh City, Vietnam. A rain gauge, satellite communication antenna, and solar panels are located on the roof.

Latest in Forecasting Operations

Modern geophysical science has convincingly demonstrated that to predict the behavior of weather and climate processes and, in particular, severe and high-impact weather events more than a few days in advance, it is essential to have access to data from the entire globe—from within the upper layers of the ocean and land surfaces, as well as from the atmosphere. That access permits scientifically based predictions of the individual synoptic-scale weather systems, such as tropical cyclones and anticyclones for various notional lead times (Zillman 2003a).

In the past few decades, we have witnessed major advances in the observation, analysis, and prediction of high-impact weather and climate (NRC 2007, 2008; Shapiro et al. 2010). Notable improvements have occurred in monitoring and predicting short-term high-impact weather and weather hazards, climate variability, and climate change. The accuracy of global five-day forecasts is comparable with that of two-day forecasts 25 years ago (Shapiro et al. 2010).[15] Currently, 12- to 48-hour forecasts, on spatial scales of a few kilometers, can provide timely and accurate warnings of flooding, rainstorms, river flows, tornadoes, storm surges, tropical cyclone track and landfall, and air-quality emergencies (Shapiro et al. 2010). This information is also useful in managing electricity generation, water resources, and transportation where routine weather from a meteorological perspective can have a high impact—excess heating resulting in electrical load shedding or additional high-cost generation and pollution, for example. There is also more evidence of predictability of some extreme weather events 7–10 days in advance. Significant advances have also been made on longer time-scales, and a greater understanding of forecast uncertainties permits more useful seasonal predictions (Palmer 2004), which in turn improve the predictions on shorter time-scales (see box 3.9 technical insight).

Forecasting of smaller-scale phenomena, such as thunderstorms and heavy precipitation (often related to flash flooding), within the framework of the synoptic-scale predictions requires much more detailed monitoring and modeling of the local region on generally shorter time-scales.[16] Hence, in any one country, a component of the observation network is part of the large-scale observation system, consisting of synoptic meteorological stations, upper-air measurements (see box 3.10 technical insight), ocean measurements, and a finer-scale network (mesoscale), which may consist of a combination of surface meteorological stations, simple rain gauges, and weather radar (see, for example, NRC 2009).

By sharing critical observational data through WIS, NMHSs enable global modeling centers[17] to develop NWPs (see box 3.11 technical insight), which can be shared with NMHSs that have limited numerical modeling capability to provide forecasting guidance. In some cases, these model products from the global centers provide the input for higher-resolution local area models, which use the additional data available to the NMS nationally and locally. As the resolution of global models increases, so does the value of using these tools for national and local applications, thereby resulting in a shift in investment nationally from local model development to global and regional model use.

Box 3.9 Technical Insight: How Weather and Climate Interact

Although weather describes the short-term behavior of the atmosphere and climate describes the long-term average of the weather, no physical boundary separates them; the physical processes of the atmosphere, ocean, and land operate on a continuum in time and space. In reality, weather affects climate, and climate affects weather. Consider the following scientific insights:

- Natural climate variations, such as the El Niño–Southern Oscillation and the North Atlantic Oscillation, significantly alter the intensity, track, and frequency of hazardous weather, such as extratropical and tropical cyclones and associated high-impact weather.
- No longer-term regional variations exist, such as decadal variability in tropical cyclones and multidecadal drought in the Sahel region in Africa.
- Conversely, small-scale processes have significant upscale effects on the evolution of large-scale circulation and the interactions among the components of the global system.

For scientists, one big challenge is to better predict the spatial temporal continuum of the interactions between weather and climate, thereby bridging the gap between forecasting high-impact events at daily to seasonal time-scales. Another is to develop an integrated approach to improve how we predict this Earth system without making arbitrary distinctions that have no physical basis.

Sources: Data from Brunet et al. 2010; Hurrell et al. 2009; Shapiro et al. 2010.

Box 3.10 Technical Insight: Using Radiosondes for Upper-Air Measurements

Upper-air data are particularly useful and should ideally be obtained twice per day per station. The temperature, humidity, and wind profile is obtained by a radiosonde, which consists of an instrument package and balloon (hydrogen or helium filled). The balloon ascends to a height of about 20 kilometers. The instrument package is not recoverable, and the cost of each ascent is approximately US$280. Including the cost of a ground station (with a five-year lifetime), the annual indicative operating costs of running an upper-air station are about US$210,000 (Gray 2012). Because of the high cost of expendables, many developing countries struggle to maintain their upper-air networks without external support.

Numerical Weather Prediction (NWP) provides the basic guidance for weather forecasting beyond the first few hours (Kalnay 2003). Yet the human forecaster still has a critical role in interpreting the output and in reconciling sometimes conflicting information from different sources. This role is especially important in situations of locally severe weather. The capacity to use this information depends on access to good-quality interactive workstations for overlaying and manipulating the basic information (photo 3.3).

Weather forecasting offices produce a variety of forecast products, depending on their capability and users' requirements. Many provide forecasts from 12 hours to seven days or longer. In modern NMHSs, forecasters spend most of

Box 3.11 Technical Insight: Modern Forecasting Techniques

Forecasts for lead times in excess of several hours are essentially based almost entirely on numerical weather predictions (NWPs). Weather systems several times the scale of the three-dimensional grids used in modeling can be resolved. As the grid size becomes smaller, the ability to resolve phenomena on smaller scales increases. Phenomena smaller than the grid size are represented in an approximate way using statistics and other techniques. These limitations in NWP models affect detailed forecasts of local weather events, such as intense precipitation, fog, and peak wind gusts. They also contribute to the uncertainties that can grow chaotically, ultimately limiting predictability.

The use of ensemble prediction systems (EPSs) has enabled forecasters to execute a group of forecasts—an ensemble—from a range of modestly different initial conditions or from a collection of NWP models (a multimodel ensemble) with different but equally plausible approximations. If the ensemble is well designed, its forecasts will span the range of likely outcomes, providing a range of patterns where uncertainties may grow. Operational services use EPSs extensively for weather and climate prediction because they offer an estimate of the most probable future state of the system and an estimate of a range of possible outcomes. The latter is particularly useful from the viewpoint of decision making, where minimizing risk is more often related to the quantitative estimates of occurrence of high-impact events than with the most probable future state (Rogers et al. 2010). From this set of forecasts, information on probabilities derived automatically can be tailored to users' needs.

Clearly, advances in predictability must be intrinsically linked with the applications for which the forecasts are made (Palmer 2004)—in turn linking the forecast production system directly to service delivery.

Photo 3.3 Modern forecaster workstation capable of blending observation and model products. Department of Meteorology, Ministry of Water Resources and Meteorology, Cambodia.

their time on zero- to six-hour nowcasting (see box 3.12 technical insight), with longer-period predictions relying on forecasts that use the large-scale, multimodel ensembles constructed by the major prediction centers (Mass 2012; see box 3.13 technical insight). Incorporated into a multihazard warning system, this arrangement strengthens the connection between NMHSs and disaster management and civil protection and can exploit the latest technologies to translate and communicate information about the impact of severe weather events to people in ways they understand (see chapter 4).

Box 3.12 Technical Insight: What Is Nowcasting?

Forecasts that extend from 0 to 6 or 12 hours are based on a more observation-intensive approach and are referred to as *nowcasts* (Gray 2012; Mass 2012). The strength of nowcasting lies in providing location-specific forecasts of storm initiation, growth, movement, and dissipation, which allows for specific preparation for certain weather events by people in a specific location.

It is anticipated that in a modern national meteorological service, forecasters will spend most of their time on zero- to six-hour nowcasting, with longer-period predictions relying on objective model-based forecasts that use the large-scale multimodel ensembles constructed by the major prediction centers. Forecasters will increasingly issue nowcasts of the current weather and how the weather system will evolve over the next few hours. In this time range, it is possible to forecast small features such as thunderstorms and tornadoes, which cause flash floods, lightning strikes, and destructive winds. Nowcasting is also used for aviation weather forecasts in the terminal area and en route environment, marine safety, water and power management, offshore oil drilling, construction industry, and leisure industry.

Traditionally, nowcasting focused on the extrapolation of observed meteorological fields. Nowadays, techniques combine extrapolation with numerical weather prediction (NWP) through blending and assimilating detailed observations, including radar and satellite data); radar data pick out the size, shape, intensity, speed, and direction of movement of individual storms on a practically continuous basis. The intensity of rainfall from a particular cloud or group of clouds can be estimated, which gives a good indication of whether to expect flooding or river swelling. Depending on the area of built-up space, drainage, and land use in general, a forecast warning tailored to the safety and well-being of a particular community may be issued. By including NWPs, nowcasting can be extended to produce a short-period forecast.

Nowcasting packages, developed by the EUMETSAT (European Organisation for the Exploitation of Meteorological Satellites) Network of Satellite Application Facilities, for example, use Meteosat Second Generation's SEVIRI (Spinning Enhanced Visible and Infrared Imager) and the U.S. National Oceanic and Atmospheric Administration and EUMETSAT Polar System's AVHRR (Advanced Very High Resolution Radiometer) to support operational centers' need for severe weather information on very short time-scales. Similar packages are available from other meteorological satellite operators and are used routinely as a part of the World Meteorological Organization's (WMO) Severe Weather Forecasting Demonstration Project (SWFDP). These techniques are particularly valuable where ground-based radar data are unavailable, and they provide opportunities for more rapid technological advance.

Box 3.13 Technical Insight: The Uncertainties Surrounding Long-Range Forecasting

Beyond two weeks, weekly average predictions of detailed weather events have very low skill,[a] but forecasts of one-month averages, using numerical weather prediction (NWP) with predicted sea-surface temperature anomalies, still have significant skill for some regions and seasons to a range of a few months. This predictability exists because of the inherent inertia and intrinsic time-scales of variability in the temperatures of the land surface and sea, as well as the slow changes in ocean waves and currents compared with the atmosphere. To the extent that the atmosphere is connected to the ocean and land-surface conditions, a degree of predictability may be imparted to the atmosphere at seasonal time-scales. The most important example of this coupling is the El Niño–Southern Oscillation phenomenon, which produces large swings in the climate at intervals ranging from two to seven years.

Seasonal predictability must be understood in probabilistic terms. It is not the exact sequence of the weather that has predictability at long lead times, but rather some aspects of the statistics of the weather. Although the weather on any given day is entirely uncertain at long lead times, the persistent influence of the slowly evolving surface conditions may change the odds for a particular type of weather occurring on that day. Currently, seasonal predictions are made using both statistical schemes and dynamical models. The statistical approach seeks to find recurring patterns in the climate associated with a particular predictor field (such as the sea-surface temperature). The basic tools for dynamical prediction are coupled models of the ocean, atmosphere, and land. Uncertainty in the dynamical models is handled using ensemble techniques, where the climate model is run many times with slightly different initial conditions. From this information, the statistics of the climate can be estimated.

At longer time-scales, predicting the future climate requires physically based models that represent feedback, such as cloud radiation, water vapor, ocean dynamics, sea ice, aerosols, and ocean heat transport. The treatment of these key features is adequate for reproducing many aspects of the climate realistically. However, systematic errors are still apparent—for example, in the simulated temperature distributions in different regions of the world or in different parts of the atmosphere, in precipitation fields, and in clouds (in particular, marine stratus).

The application of ensemble prediction to seasonal and longer time-scale predictions (including climate change) has also been demonstrated and is being further developed for operational use. For example, the European Union's ENSEMBLES project has developed an ensemble prediction system for climate change using high-resolution global and regional Earth system models (Van der Linden 2008). Seasonal forecasts derived from ENSEMBLES are produced at the Southern African Development Community's Climate Services Centre and are made available to the health sector for malaria control (Rogers et al. 2010).

Good observations of the Earth system at all scales are essential for improving the quality of these models and for developing observed climatologies that can be used in infrastructure planning.

a. *Skill* in this context refers to the representation of error that relates to the accuracy of a particular forecasting model.

Because flooding is often one of the predominant hazards in many places, most NMHSs need access to tools such as Doppler radar, which has proved to be the most valuable tool in detecting high-impact weather. Assimilating radar and satellite data into skillful[18] high-resolution local area models provides the forecaster with the ability to warn on forecast rather than to warn on detection. This approach has the potential to provide longer lead times for severe forecasts from the current realizable limit of about 20 minutes for warn-on-detection approaches, thereby providing emergency managers with much earlier warnings of hazardous weather and more time to make effective decisions. The technological steps required to achieve this level of skill are challenging for many developing countries but are not impossible.

Limiting Factors in Forecasting

Support for NMHSs observation networks—despite their importance for detecting high-impact weather events—has proved to be one of the biggest challenges and a limiting factor in the effectiveness of forecasting and warning systems. Gray (2012) reports that although developing countries have put much effort into strengthening land-based weather station networks for synoptic and mesoscale observations, their success is often limited.

One key problem is that the focus of projects is usually on the infrastructure of the weather stations rather than on their outputs and how those outputs are used in disaster risk reduction and other services. Consequently, weather stations are often successfully installed, but staff members receive a limited amount of training in maintaining them. Ultimately, the station ceases to function, owing to lack of funding for spare parts, lack of integration into the existing observing network, or lack of trained staff members. Without the capacity to demonstrate the benefit of the new stations, funding is not provided to enhance the service.

In particular, there is a paucity of data throughout Africa from synoptic stations, which inevitably results in poorer-quality numerical guidance and forecasts in those regions. Calibration of sensors used in surface observations is very important (Gray 2012), but in practice, few sensors are calibrated to internationally accepted standards.

Vertical profiles of temperature and humidity from radiosondes are a high priority because these data are important for monitoring climate and for assimilating in NWP models. Operational upper-air stations are particularly sparse in Africa,[19] primarily because of a lack of consumables. The WMO has estimated that, in Africa alone, there is a need for an additional 4,000–5,000 basic meteorological observations. As an ideal target, the Regional Basic Synoptic Network should have a horizontal resolution of 150 kilometers for surface observations (WMO 2011). Although several initiatives have taken up this issue, the network still has large gaps.

One approach is to explore public-private partnerships, which could support new observations that would be financed and used by the private sector to meet its needs, while providing these data to the NMSs to enhance their capacity to provide better public forecasts and warnings. A large-scale example of this approach is the Oklahoma Mesonet in the United States (see box 3.19 technical insight, later in this chapter), and the U.S. National Research Council envisions a distributed adaptive *network of networks*—that is, a network of multiple environmental applications jointly provided by the government, industry, and the public, at least within the United States (NRC 2009).[20]

Another key problem is that despite considerable progress in observation and forecasting, uncertainties in forecasts will always remain. There is a risk that the public will come to expect the forecasts to always be right, and when they are wrong, they must be the result of incompetence, negligence, or some other culpable form of system failure (WMO 2002). What should be understood is that some meteorological phenomena will remain inherently unpredictable, and the more extreme the phenomena, the more likely this will be so.

Best Practices in Service Delivery

After the data are collected and analyzed, they need to be delivered, in an appropriate form, to the various users. For the public, this information can be delivered quickly through conventional media, as well as increasingly through mobile devices and the Internet. If the customer is a civil protection agency, the information will be in the form of an early indicator of the likely risk of a potential threat to lives and livelihoods and may be delivered ahead of any information conveyed to the public. If the user is a weather-sensitive sector, such as electricity generation, information will be adjusted to the different needs of grid operators, power supply and maintenance crews, and transportation networks. The process of interpretation, delivery, and communication with end users is often referred to as the *Public Weather Service Platform* (see photo 3.4).

Skills are needed in understanding both the implications of the forecast and the specific uses of the forecast information. The role is one of transforming and translating scientific information into something that is actionable by different parts of society. This role is often referred to as *impact forecasting*—that is, forecasting the impact of the weather rather than just the weather. For example, a forecast of 50 millimeters of rain in the next hour is less useful than a forecast that explains that 50 millimeters of rainfall might result in flooding of homes located in a vulnerable location. This task is unfamiliar in many NMHSs, particularly in developing countries struggling to support their basic observation and forecasting infrastructure. But if it is not emphasized, the capacity to influence government and societal decisions is severely impaired, which in turn contributes to the organization's declining influence and budget support. A frequent criticism from end users of NMHSs, particularly in developing countries, is that they do not understand the information given and little care has been put into making the weather forecast useful for them (see box 3.14). More effective partnerships between providers and the main users of weather, climate, and water information are essential.

Photo 3.4 Public Weather Service delivery system, (Meteorological and Geophysical Agency, Indonesia).

Source: Photo courtesy of H. Kootval.

The WMO is addressing this communication problem by implementing a strategy for service delivery, which is intended to serve as a foundation for improving services by sharing best practices and guidelines, along with increasing user engagement (WMO 2012b). The WMO is encouraging NMHSs to develop more impact-based forecast and warning services. Table 3.1 shows examples of the language adopted in Croatia.

Weather forecasts and warnings should be demand driven, impact based, and tailored to user-defined thresholds. Such forecasts will ensure that critical weather information is communicated about societal impacts to individuals and sectors most at risk. This information should be made available to the community in a variety of easy-to-understand formats. The attributes of service delivery focus on the needs of users (see box 3.15) and aim to ensure that forecast and warning production is equally matched by appropriate responses.

As an example of how weather information can be communicated to the public and specialized users on behalf of the WMO, the Hong Kong Observatory has developed the Severe Weather Information Centre (SWIC).[21] SWIC aims to make local severe weather warnings issued by official weather services (NMSs) available to all interested individuals globally. SWIC was adopted as the dissemination platform with the support of the WMO Public Weather Service Programme[22] and was implemented by Hong Kong SAR, China. Local warnings issued in Guam; Hong Kong SAR, China; the Republic of Korea; Macao SAR, China; and Singapore are disseminated through a widget on desktop or laptop computers known as SWIdget.[23]

Box 3.14 Kenyan Farmers Need Better Forecasts

In a recent survey of weather and climate information needs of small-scale farming and fishing communities in Kenya, 74 percent of the 401 respondents rated the weather and climate information they were receiving as only somewhat useful to their operational decisions (Awiti, Onyango, and Ochieng 2012). The reason for this response is that scientific weather and climate forecasts were formulated on a much wider geographic scale than could be used by farmers and were often presented in a language and format unfamiliar to them. Furthermore, the information did not address their specific needs, which related to the onset and cessation of rain, drought forecasts, and pests and disease outbreak. Nor did the information address the needs of fishers for hourly forecasts on winds and currents.

This mismatch between the forecast and the end users' needs can undermine trust in the institution providing the forecast, as well as its application by farmers and fishers to their operational decisions. Weather and climate forecasts must address users' needs; the forecaster must understand the decision options that are sensitive to end users' decision-making objectives and constraints; and the forecast must be delivered at the right time and at an appropriate scale, with sufficient accuracy for relevant decisions.

Moreover, the study found that the forecast is typically presented in jargon and is incomprehensible to a large majority of smallholder farmers and fishers, who have no more than a primary school education. For example, if the forecast is "rainfall slightly enhanced in March, near normal in April, and slightly depressed in May," farmers would like to know the implications for maize production and management with respect to planting time, seed variety, fertilizer application, rainfall distribution, and projected yield.

The World Bank is beginning to address this issue with programs such as a modernization of the Nepal Department of Hydrology and Meteorology funded by the Pilot Program for Climate Resilience in Nepal and the introduction of an agricultural management information system. That system should ensure that the problem of interpreting meteorological outputs is overcome before information reaches the farmer.

Table 3.1 Adoption of More User-Friendly Forecast Language in Croatia

Before	After
There is a 60% chance of thunderstorms this afternoon.	Thunderstorms between 2:00 and 4:00 p.m. will likely cause 30- to 60-minute flight delays.
Thunderstorms will be in the response area this afternoon.	Responders should seek shelter owing to the possibility of lightning from 2:00 to 4:00 p.m.
Heavy snow will fall with accumulations of 8–12 inches tonight.	Interstate Highway 80 will likely become impassable after midnight because of heavy snowfall.
Warning: sea state 4.	Moderate sea, with waves reaching a height of 1.25–2.5 meters, will likely present danger to smaller vessels such as speed boats and excursion boats.
Warning: gusts of Bora wind will reach over 100 kilometers an hour in Makarska region.	Gusts of Bora wind will likely cause an interruption in maritime traffic, especially the Makarska-Sumartin (Brač Island) ferry line.

Source: Data from WMO.

Box 3.15 Attributes of Service Delivery Defined by the World Meteorological Organization

The World Meteorological Organization (WMO) has defined the following attributes for the delivery of forecasts:

- *Available and timely.* Service delivery should occur at time- and space-scales that the user needs.
- *Dependable and reliable.* Delivery should be on time to the required user specification.
- *Usable.* Information should be presented in user-specific formats that the client can fully understand.
- *Useful.* Information should respond appropriately to user needs.
- *Credible.* The user should be able to confidently apply the information to decision making.
- *Authentic.* Information should be entitled to be accepted by stakeholders in the given decision contexts.
- *Responsive and flexible.* Information should meet to the user's evolving needs.
- *Sustainable:* Service should be affordable and consistent over time.
- *Expandable:* Information should be applicable to different kinds of services.

This strategy identifies four stages in the delivery process:

- *Stage 1: User engagement.* Identifying users and understanding their needs, as well as understanding the role of weather, climate, and water information in different user sectors.
- *Stage 2: Service design and development.* Creating, designing, and developing services to ensure users' needs are met.
- *Stage 3: Delivery.* Producing, disseminating, and communicating data, products, and information (such as services) that are fit for their purpose and relevant to users' needs.
- *Stage 4: Evaluation and improvement.* Collecting user feedback and performance metrics to continuously evaluate and improve products and services.

Source: WMO Public Weather Service Programme.

A similar system, the World Weather Information Service, has been developed for official and authoritative weather forecasts by NMHSs from about 130 countries and territories. A website, run by the Hong Kong Observatory on behalf of the WMO, is available to anyone, anywhere, and anytime. A mobile version, MyWorldWeather,[24] was launched in October 2011 for people on the move. This application also provides official weather forecasts for nearby cities.

A New Focus on Training

NMHSs require staff members to keep their professional skills up to date and learn how to communicate with a variety of stakeholders (see box 3.16). The WMO provides guidelines for qualifications of meteorological and hydrological personnel through the WMO Education and Training Office.[25] Although the

> **Box 3.16 Teaching Meteorologists How to Communicate with Economists and Sociologists**
>
> As early as 1968, the meteorological community recognized the importance of meteorology for economic development, with the First International Seminar on the Role of Meteorological Services in Economic Development in Africa (WMO 2003). This application posed a new problem, however, because, unlike the mariner or aviator, the development planner is unlikely to be knowledgeable about meteorology—and much less certain of its value to his or her decisions.
>
> Addressing this issue, Étienne Bernard (1976) compiled lecture notes on training personnel to apply meteorology to economic and social development. At that time, he was optimistic that governments and economic planners would call on meteorologists more frequently to address food shortages, water requirements, energy needs, and environmental pollution. He drew attention to the need for meteorologists to be acquainted with social and economic factors and to be able to express themselves in appropriate economic terminology. He noted that any dialogue between economists, development planners, and other competent authorities would be impossible unless the meteorologist was able to speak their language. More than 40 years later, this process is still in its infancy. Further progress has been made following the World Meteorological Organization's (WMO) Madrid Conference on the Social and Economic Benefits of Weather, Climate, and Water Information and Services in 2007, which led to the Madrid Action Plan (WMO 2007) and subsequent activities by different WMO regional associations and programs.

basic qualifications are usually met by university courses, WMO Regional Training Centers (RTCs) offer specialized courses and workshops in areas such as (a) the latest developments on the use and interpretation of NWP models, (b) competence assessment of aviation forecasters and observers, and (c) radar meteorology. Distance learning is now readily available so that the cost of training large numbers of students can be minimized. However, face-to-face workshops are still essential elements of meteorological and hydrological training. A European virtual organization for meteorological training, Eumetcal, was set up by Eumetnet, a grouping of 29 NMSs, to share training materials and exploit e-learning programs and methods. Eumetcal brings together tools from many different sources, including EuMeTrain, the U.K. Met Office, the European Centre for Medium-Range Weather Forecasts, EUMETSAT, and the COMET Program.

In several countries, synergies exist among universities, research centers, and NMHSs:

- In France, the research department of Météo France is cohosted by a research institution (the National Center for Scientific Research), ensuring that Météo France is well connected with the research community.
- In the United Kingdom, the Met Office has strong ties with universities (such as Reading and Exeter) and operates the Met Office College, which provides training for its own staff, as well as the staff of other NMHSs.

- In the United States, universities and the National Oceanic and Atmospheric Administration are also complementary (see box 3.17).

With regard to training and staffing, creating university degrees is an important step in developing well-functioning NMHSs. In China, a WMO training center is hosted by Nanjing University in collaboration with the China Meteorological Administration, which offers degree programs for international and national students, as well as short technical courses. These university degrees are cosponsored by an NMS and a university to ensure that the training fits the needs of the NMS. In China, this arrangement is part of its international development assistance. Roshydromet has a similar relationship with universities in the Russian Federation, especially the Russian State Hydrometeorological University, as well as its own Roshydromet Advanced Training Institute.

Staff members of NMHSs are also required to understand users' needs and work closely with users. The importance of collaboration between producers and users of weather, climate, and water products and services cannot be overestimated (Rogers et al. 2007). This cooperation ensures that value is added where it is needed so that the user properly considers and acts on environmental information.

Aeronautical meteorology, marine meteorology, and agricultural meteorology are obvious examples of fields that have developed within the framework of this close relationship. In each of these cases, the decision maker is a knowledgeable user of weather information, usually with some training in meteorology or hydrology because of the high risk to the activity from adverse weather. New services

---

Box 3.17 U.S. Efforts to Broaden Training in the Environmental Sciences

The COMET Program was established in 1989 by two U.S. agencies—the National Oceanic and Atmospheric Administration's, the National Weather Service, and the University Corporation for Atmospheric Research—to promote a better understanding of mesoscale meteorology among weather forecasters and to maximize the benefits of new technologies during the National Weather Service's modernization program. Its mission has now expanded to include the use of innovative methods to disseminate and enhance scientific knowledge in the environmental sciences—particularly meteorology, but also oceanography, hydrology, space weather, and emergency management.

COMET's MetEd website (http://comet.ucar.edu) provides education and training resources to benefit the operational forecaster community; university atmospheric scientists and students; and others interested in learning about meteorology, weather forecasting, and related geoscience topics. MetEd is supported by U.S. institutions, as well as by the Australian Bureau of Meteorology, the European Organisation for the Exploitation of Meteorological Satellites (EUMETSAT), and the Meteorological Service of Canada.

are frequently developed, tested, and implemented with the direct cooperation of users through a well-defined requirements process.

The greater emphasis on decision support requires capabilities that differ from those traditionally found in NMHSs. Cross-sectoral training is needed to increase the capacity and capability of the producer and consumer of hydrometeorological information to work together. For example, WMO cooperating organizations, such as the International Research Institute for Climate and Society, are now teaching courses aimed at employees of health, climate, hydrometeorological, and related services.[26] This new capability, at the interface of hydrometeorological and user sectors, will lead to better decision tools and more effective outcomes for society and the economy.

This type of training is relatively new in sectors such as energy, health, and planning, but it is more common in agriculture, aviation, and marine transportation. There are opportunities to find best practices in current training efforts that may be universally applicable to cross-sectoral training (Rogers et al. 2007).

A service-oriented or customer focus implies good communication skills. The trend in many advanced NMHSs is to trade the cost of increased technological innovation for a decrease in staffing, often closing offices and consolidating capacity in central facilities. This approach tends to reinforce the ability to provide forecast products, but it may reduce the capacity of NMHSs to work directly with users. In this case, the capability to exploit modern information technologies is essential.

Those NMHSs that effectively embed their employees within their key customer organizations generally develop a more effective producer and consumer relationship, which leads to greater innovation in the customer's sector and to closer alignment of the NMHSs with the expected outcomes of that provided service. Many examples of this approach exist, including road weather services, support for military operations, agricultural services, and airport operations (see box 3.18).

More emphasis has also been placed on the social and economic evaluation of weather, climate, and water products and services requiring different expertise than traditionally found in NMHSs. In this area, there is still considerable room for improvement and an opportunity to transfer skills from institutions such as the World Bank.

Creating New Partnerships

Advances in science and technology are driving the evolution of the weather and climate enterprise. In the past, government agencies collected nearly all of the weather, climate, and hydrological data and ran nearly all of the forecast models (NRC 2003). Today, local agencies, universities, and private companies can deploy their own instruments—some run their own models or models developed by others—and provide services to specific clients. These advances will continue and will further increase the strain among the various actors. For example, tensions exist between some meteorological departments

Box 3.18 On-Site Delivery of Localized Forecasts at Heathrow Airport

Heathrow is the world's busiest international airport. With more than 476,000 aircraft movements and 69 million passengers traveling through it every year, any small delay can cause ripple effects, slowing the carefully planned logistics of transiting many people through the airport.

The U.K. Met Office has provided services to Heathrow for some time. Although it can provide tailored services remotely, Heathrow invited the Met Office to work on site, side by side with its operational staff—around the clock annually. Consequently, the Met Office quickly gained an understanding of airport operations and thresholds that would have been impossible from a different location.

Heathrow staff members can call in at any time to ask forecasters for the latest information. They can discuss aspects that are most important to them, see the graphics that the Met Office uses to forecast, and discuss the probabilities attached to any risk.

According to Jon Proudlove, general manager for advanced transport systems at Heathrow,

> The weather can impact significantly on the Heathrow operation. Having the Met Office forecaster embedded within the Operational Efficiency Cell at the Control Tower is without doubt bringing benefit to the ATC [air traffic control] operational decision making process. Our learning around the timely sharing of information continues to develop and I can only see the benefit of the co-located forecaster increasing. For example the decision making around Runway Changes, low visibility operations, and the prediction of snow events have all been enhanced.

Source: U.K. Met Office, http://www.metoffice.gov.uk/aviation/heathrow.

---

in Africa and private agricultural businesses that need weather and climate information that is fit for purpose of offering insurance and other services to their clients.

This area also has considerable capacity within private weather service providers and is a potential source of tension between the public and private sector, which may be perceived to be in direct competition. Access to basic meteorological information by the private commercial sector and by NMHSs competing for the same value-added services is an inherent problem that is not easily solved without a regulatory framework that creates a so-called level playing field, which charges all actors providing commercial services the same for access to the same information, be they commercial companies or NMSs providing commercial services. The perceived advantage of NMHSs, absent a strong regulatory framework for data sharing, creates many of the difficulties encountered between the public and the private sectors. Some countries (such as the United States) restrict the activities of NMHSs to prevent them from competing with the private sector, whereas others require NMHSs to offer commercial services and rely on competition laws to engender fair competition between the public and the private sectors (see chapter 5).

In the absence of government-supported networks or in parallel, the private sector and academia are creating their own private observation networks and using and sharing these data. These separate networks may eventually provide more capacity to the national observation network than the NMHSs could do alone, assuming they are developed cooperatively and with an understanding of the limitations and utility of each of the observation systems. Rather than pursuing an adversarial position, which is often the case, government and the private sector need to find ways for them both to benefit. This objective could be achieved through effective national coordination mechanisms. An important role for the NMHSs in this environment is to define observation system standards that are based on WMO guidelines, to adhere to them, and to provide authoritative information that can serve as a benchmark for other observation networks. An example of an effective partnership is the meteorological observation network in Oklahoma (see box 3.19 technical insight).

Box 3.19 Technical Insight: Oklahoma Mesonet

The Oklahoma Mesonet is a world-class high-resolution network of environmental monitoring stations developed and implemented by two universities: the University of Oklahoma and Oklahoma State University. The network complements and extends the National Weather Service's surface observation network, providing the high-resolution measurements needed for many local client applications, including public safety, agriculture, wildfire management, and K–12 education. This nonadversarial approach maintains the national meteorological service's natural monopoly on the networks required for production of forecasts and warnings of severe weather, but it provides useful additional data to improve the accuracy and timeliness of local warnings.

The Oklahoma Mesonet consists of 120 automated weather stations, which provide data for the Oklahoma Climatological Survey. Each of Oklahoma's 77 counties has at least one station. At each site, the environment is measured by a set of instruments located on or near a 10-meter-tall tower. The measurements are packaged into observations every five minutes and then transmitted to a central facility every five minutes, 24 hours per day year-round. The Oklahoma Climatological Survey at the University of Oklahoma receives the observations, verifies the quality of the data, and provides the data to Mesonet customers. It takes only 5–10 minutes from the time the measurements are acquired until they become publicly available.

OK-First is an outreach project of the Oklahoma Climatological Survey and Oklahoma Mesonet. It provides training and real-time weather data to public safety officials for use in weather-impacted situations. OK-First training and data are provided at no cost to qualified applicants in Oklahoma. As of March 2011, more than 190 agencies in and around Oklahoma were participating in the program, including the local weather forecast offices of the National Weather Service. OK-First receives substantial funding from the Oklahoma Department of Public Safety. Costs are recovered according to a well-defined data policy.

Top Priorities for Improving NMHSs

Modernization

Modernization is a major issue for NMHSs in both developing and developed countries. Some 88 percent of the respondents to the 2011 WMO survey indicated that modernization was second in importance only to the level of government funding (WMO 2012a). Much of the technical infrastructure (observation systems, communications facilities, data archival systems, and forecasting systems) of developing countries is obsolete. Although modernization programs are now under way in several countries, considerable effort is needed to strengthen most developing countries' NMHSs.

Staffing and Government Funding

The absence of a critical mass of expert staff members and funding to plan, implement, operate, and maintain state-of-the-art equipment and systems is another barrier to progress (see chapter 6). Most countries lack the government funds for equipment replacement. Even NMHSs in developed countries lack the funding to replace their basic infrastructure on reasonable time-scales, and they have limited capacity to introduce new systems and technologies emerging from the international research effort. Countries that have made and sustain these investments, such as China and Korea, have rapidly brought their services to a very high standard.

Capacity to Exploit Scientific Capability

Although considerable progress has been made in modeling weather and climate, the scientific expertise required to develop and run these models, as well as the necessary computing resources and infrastructure, means that advanced operational modeling remains the preserve of a few developed NMSs (Gray 2012). The additional demands for convective-scale models, ensemble prediction systems (EPS), and more sophisticated Earth system models mean that even developed NMSs and global and regional modeling centers struggle to access the computing power needed to produce NWP forecasts and climate predictions to the full potential of the scientific capability (Shapiro et al. 2010).

Global and Regional Support Centers

NMHSs depend on the international network of observations, numerical weather and climate predictions, and—where available—operational assistance from WMO regional centers. The WMO's development of Severe Weather Forecasting Demonstration Project (SWFDP) is demonstrating the importance of these regional centers in increasing the capability of NMHSs to deliver services nationally (see box 3.20 technical insight). However, this regional support cannot be easily sustained without targeted investment. Ideally, each of the benefiting NMSs should contribute financially to the operations of a regional center, but often they are unable to do so because of their own budgetary constraints. Thus, programs aimed at strengthening NMHSs should also consider the needs of

Box 3.20 Technical Insight: Severe Weather Forecasting Demonstration Projects

The World Meteorological Organization (WMO) has developed and implemented a series of Severe Weather Forecasting Demonstration Projects (SWFDPs), which are designed to demonstrate the usefulness of global numerical weather prediction (NWP) products, particularly ensemble prediction systems (EPS)—produced by global meteorological centers and regional specialized meteorological centers (RSMCs) in improving severe weather forecasting services in countries where sophisticated model outputs are currently not used.

These projects directly link improvements in NWP with improvements in forecasting skill at the national level and with improvements in services delivered to stakeholders. Currently, SWFDPs are running in Southern Africa and in the South Pacific Islands, with new projects in development for Southeast Asia, Central Asia, and Eastern Africa.

RSMC Pretoria, operated by the South African Weather Service, continues to support the SWFDP in Southern Africa. RSMC Pretoria has helped NMSs improve their use of global modeling products (a) to predict severe weather events, such as heavy rain and strong winds, (b) to improve alerts, (c) to improve interaction between NMSs and civil protection, and (d) to improve the global data-processing and forecasting system (GDPFS) and service delivery overall. Similar activities are under way in most of the WMO regions, using RSMCs to connect with the global product centers (such as the European Centre for Medium-Range Weather Forecasts, the U.K. Met Office, and the U.S. National Weather Service's National Centers for Environmental Prediction) to support and assist participating NMSs.

WMO regional specialized meteorological centers (RSMCs) and their role in supporting NMHSs.

Public Weather Services

Public weather services are a major issue for many NMSs, which continue to rely on providing weather products rather than on developing services that match users' needs.[27] The growing recognition of the utility of weather and climate information for key social and economic decisions means that public weather services are expected to be the platform for warning information that is tailored to the specific requirements of all sectors—from civil protection to the general public—in a form that is readily usable.

Competition

In many countries, the public weather service function can also be an area of intense competition between (a) the NMS and (b) the entrepreneurial media sector, other NMHSs, and the private sector. Such competition raises the issue of the boundaries between freely available basic services and those appropriate for cost recovery. The development of observation networks separate from NMHSs creates a dilemma: more data are available, but access requires the NMHSs to partner with other agencies and the private sector.

Provision of Climate Services

Many NMSs have operated as weather services, playing only a small role in archiving climate data and none in providing climate services to the general public or to specialized users. There is a risk that these services will be poorly placed to play a significant role in the growing need for climate services and within the new Global Framework for Climate Services (WMO 2009b) and national climate services (Zillman 2011).

National Role of NMHSs

Government support of the NMHSs as the single official voice for warning services at the national level is an issue in many countries. Absent a legal and regulatory framework, self-appointment and de facto recognition by the WMO is insufficient to ensure that NMHSs have the necessary authority to provide official warnings. This issue also touches on the responsibilities of other government departments, the media, and civil society to take appropriate action. Poor positioning of NMHSs in the government structure often results in budget support being a low priority. This circumstance produces a cycle of decline owing to inadequate funding of operations and infrastructure, and it results in poor services and further government neglect (Zillman 2005).

In chapter 4, the value of an intragovernment approach to warning services is discussed. By integrating meteorological and hydrological warnings into a multihazard warning system, earlier alerts, warnings, and action are possible. In this way, the critical role of NMHSs becomes more apparent to the government and central to warning services.

Notes

1. The tag *weather, climate, and water* is used frequently instead of meteorology and hydrology. Meteorology is inclusive of weather and climate and these terms are interchangeable. *Water* is the tag used to refer to hydrology and occasionally oceanography.
2. For more information about World Weather Watch, see its page on the WMO website (http://www.wmo.int/pages/prog/www).
3. A very useful resource is the frequently updated *WMO Guide to Meteorological Instruments and Methods of Observation* (WMO 2008). Updated in 2010, with a corrigenda added in 2012, the guide is available at http://www.wmo.int/pages/prog/www/IMOP/CIMO-Guide.html.
4. For more information about WIGOS, see its page on the WMO website (http://www.wmo.int/pages/prog/www/wigos).
5. GOS is described on the WMO website at http://www.wmo.int/pages/prog/www/wigos.
6. GAW is described on the WMO website at http://www.wmo.int/pages/prog/arep/gaw/gaw_home_en.html.
7. WHYCOS is developed for promoting a bottom-up approach, from the country level through the basin to global scale. WHYCOS and its components focus primarily on

strengthening technical and institutional capacities of NHSs and on improving their cooperation in managing shared water resources. WHYCOS helps NHSs better fulfill their responsibilities by improving the availability, accuracy, and dissemination of water resource data and information through the development and implementation of appropriate national and regional water resource information systems, thereby facilitating the use of such systems for sustainable socioeconomic development. To learn more about WHYCOS, visit its website at http://www.whycos.org/whycos.

8. The Global Climate Observing System (GCOS) is cosponsored by the WMO; the United Nations Environment Program; the United Nations Educational, Scientific, and Cultural Organization's Intergovernmental Oceanographic Commission; and the International Council for Science.

9. For more information about the GDPFS, see the WMO website at http://www.wmo.int/pages/prog/www/DPS/gdps.html.

10. For more information about these centers, see the WMO website at http://www.wmo.int/pages/prog/www/DPS/gdps-2.html#RSMCs.

11. These services include public weather services, marine weather service, aviation weather services, agrometeorological services, climate services, environmental services, hydrological services, and health forecasting and warning services.

12. Through its advisory group on the role and operation of NMHSs, the WMO carried out a number of surveys of its members between 2000 and 2012 to understand the state and operating environment of NMSs around the world.

13. Responses are from a survey of 128 WMO members conducted during 2000–01, and a survey of between 54 and 86 members conducted from June to August 2011 (WMO 2003, 2012a; Zillman 2003b).

14. The WMO's general regulations specify that operational hydrology comprises (a) measurement of basic hydrological elements from networks of hydrological and meteorological stations; (b) hydrological forecasting; and (c) development and improvement of methods, procedures, and techniques in network design, instruments, standardization of instruments and methods of observation, data transmission and processing, supply of meteorological and hydrological data for design purposes, and hydrological forecasting (WMO 2006).

15. Unfortunately, these improvements are not universal. The skill of weather and climate prediction in the tropics continues to be limited because of the inability to realistically represent the life cycle of equatorial waves and organized convection. These inadequacies compromise the skill of forecasts on time-scales of days to weeks and beyond, including projections of climate change. Work is in hand that will improve the representation of organized convection, which in turn will improve the representation of tropical and extratropical interactions and will lead to more skillful prediction of regional-to-global weather and climate (Brunet et al. 2010; Rogers et al. 2010).

16. The boundaries between time-scales are blurring as global weather prediction models operate at finer resolution. In the near future, operational global weather prediction models will use 4-kilometer grids with nested limited area models using 1-kilometer grids, whereas climate models will run operationally at 25-kilometer or finer resolutions. The distinction between who runs what and for whom will likely change as these capabilities are introduced into the operational systems; global modeling centers will be able to provide very high-resolution products that can be used directly by NMHSs without the need for additional downscaling. It will be important, however, to ensure that as much local data as possible are assimilated

by these models, which will have implications for the way NMHSs operate in the future.

17. Global modeling centers include organizations such as the U.S. National Weather Service's National Centers for Environmental Prediction, European Centre for Medium-Range Weather Forecasts, U.K. Met Office, China Meteorological Administration, Japan Meteorological Agency, Korean Meteorological Agency, Bureau of Meteorology of Australia, Meteorological Service of Canada, Center for Weather Forecasts and Climate Studies of Brazil, Météo France, and the U.S. National Center for Atmospheric Research. These organizations are currently participating in the THORPEX (The Observing System Research and Predictability Experiment) Interactive Grand Global Ensemble in anticipation of the development of a global interactive forecasting system that contributes to the WMO Severe Weather Forecasting Demonstration Projects (SWFDPs).

18. *Skill* in this context refers to the representation of error that relates to the accuracy of a particular forecasting model. A perfect forecast results in a skill of 1.0. Forecasters apply a standard set of verification measures to compare the skill of different models. For more information, see the Centre for Australian Weather and Climate Research's website at http://www.cawcr.gov.au/projects/verification.

19. See Gray (2012) for a more detailed description of observation systems, including automatic weather stations, upper-air observations, weather radar and lightening detection, marine observations, weather satellite observations, and climate observations.

20. Application in developing countries is more difficult. Nascent efforts are under way in Kenya, the Philippines, and elsewhere to supplement national observation networks with privately supported systems; overcoming inherent tensions between government and the private sector is, however, a major obstacle. Success will depend on applying acceptable standards for observations and legal agreements on data sharing. NMHSs' lack of ability to use high-resolution observations and a perceived loss of sovereignty over data acquisition are also limiting factors.

21. SWIC's website is http://severe.worldweather.wmo.int.

22. The WMO's Public Weather Services Programme is described on the WMO website at http://www.wmo.int/pages/prog/amp/pwsp.

23. SWIdget can be downloaded at http://severe.worldweather.org/swidget/swidget.html.

24. Application is available free from mobile phone app stores. Go to https://itunes.apple.com/hk/app/myworldweather/id453654229?mt=8.

25. The Education and Training Office manages the education and training program, which serves as an advisory body on all aspects of technical and scientific education and training in meteorology and operational hydrology. Training courses are developed by the respective WMO programs with specialization in the different areas of expertise, which embraces the application of weather, climate, and water information, as well as the improvement of forecasters' and meteorological technicians' skills.

26. See the International Research Institute for Climate and Society website at http://portal.iri.columbia.edu/portal/server.pt?open=512&objID=1094&mode=2.

27. A very high priority is the provision of aviation services, which must now conform to the International Civil Aviation Organization's new provisions concerning the quality management of meteorological services in international air navigation. The provisions require upgrading quality management systems from recommended practices to international standards (that is, the International Organization for Standardizations 9,000 series of Quality Management System standards).

References

Awiti, Alex O., Silas Ochieng, and Esther Onyango. 2012. "Weather and Climate Information Needs of Small-Scale Farming and Fishing Communities in Western Kenya for Enhanced Adaptive Potential to Climate Change." *Sustainable Research and Innovation Proceedings* 4: 187–93.

Bernard, Etienne A. 1976. *Compendium of Lecture Notes for Training Personnel in the Applications of Meteorology to Economic and Social Development*. Geneva, Switzerland: World Meteorological Organization.

Brunet, Gilbert, Melvyn Shapiro, Brian Hoskins, Mitch Moncrieff, Randall Dole, George N. Kiladis, Ben Kirtman, Andrew Lorenc, Brian Mills, Rebecca Morss, Saroja Polavarapu, David Rogers, John Schaake, and Jagadish Shukla. 2010. "Collaboration of the Weather and Climate Communities to Advance Subseasonal-to-Seasonal Prediction." *Bulletin of the American Meteorological Society* 91 (10): 1397–406.

Gray, Mike. 2012. "NMS Data Availability Study for World Bank." Met Office, Exeter, U.K.

Hallegatte, Stephane. 2012. "A Cost Effective Solution to Reduce Disaster Losses in Developing Countries: Hydrometeorological Services, Early Warning, and Evacuation." Policy Research Working Paper 6058, World Bank, Washington, DC.

Hurrell, James, Gerald A. Meehl, David Bader, Thomas L. Delworth, Ben Kirtman, and Bruce Wielicki. 2009. "A Unified Modeling Approach to Climate System Prediction." *Bulletin of the American Meteorological Society* 90 (12): 1819–32.

Kalnay, Eugenia. 2003. *Atmospheric Modeling, Data Assimilation, and Predictability*. Cambridge, U.K.: Cambridge University Press.

Mass, Clifford. 2012. "Nowcasting: The Promise of New Technologies of Communication, Modeling, and Observation." *Bulletin of the American Meteorological Society* 93 (6): 797–809.

NRC (National Research Council). 2003. *Fair Weather: Effective Partnerships in Weather and Climate Services*. Washington, DC: National Academies Press.

———. 2007. *Earth Science and Applications from Space: National Imperatives for the Next Decade and Beyond*. Washington, DC: National Academies Press.

———. 2008. *Earth Observations from Space: The First 50 Years of Scientific Achievements*. Washington, DC: National Academies Press.

———. 2009. *Observing Weather and Climate from the Ground Up: A Nationwide Network of Networks*. Washington, DC: National Academies Press.

Palmer, Tim. 2004. "Progress towards Reliable and Useful Seasonal and Interannual Climate Predictions." *WMO Bulletin* 53 (4): 325–32.

Rogers, David P., S. Clark, Stephen J. Connor, Peter Dexter, Laurent Dubus, Johannes Guddal, Alexander I. Korshunov, Jeffrey K. Lazo, Marina I. Smetanina, Bruce Stewart, Tang Xu, Vladimir V. Tsirkunov, Sergey I. Ulatov, Pai-Yei Whung, and Donald A. Wilhite. 2007. "Deriving Societal and Economic Benefits from Meteorological and Hydrological Services." *WMO Bulletin* 56 (1): 15–22.

Rogers, David P., Melvyn A. Shapiro, Gilbert Brunet, Jean-Claude Cohen, Stephen J. Connor, Adama Alhassane Diallo, Wayne Elliott, Kan Haidong, Simon Hales, Debbie Hemming, Isabelle Jeanne, Murielle Lafaye, Zilhore Mumba, Nirivolona Raholijao, Fanjosa Rakotomanana, Hiwot Teka, Juli Trtanj, and Pai-Yei Whung. 2010. "Health and Climate: Opportunities." *Procedia Environmental Sciences* 1: 37–54.

Seiz, Gabriela, and Nando Foppa. 2007. "National Climate Observing System (GCOS Switzerland)." MeteoSwiss, Zurich, Switzerland.

Shapiro, Melvyn, Jagadish Shukla, Michel Beland, John Church, Kevub Trenberth, Brian Hoskins, Guy Brasseur, Mike Wallace, Gordon McBean, Antonio Busalacchi, Ghassem Asrar, David Rogers, Gilbert Brunet, Leonard Barrie, David Parsons, David Burridge, Tetsuo Nakazawa, Martin Miller, Philippe Bougeault, Richard Anthes, Zoltan Toth, Gerald Meehl, Randall Dole, Mitch Moncrieff, Herve Le Treut, Alberto Troccoli, Tim Palmer, Jochem Marotzke, John Mitchell, Adrian Simmons, Brian Mills, Oystein Hov, Haraldur Olafsson, and Jim Caughey. 2010. "An Earth-System Prediction Initiative for the 21st Century." *Bulletin of the American Meteorological Society* 91 (10): 1377–88.

Van der Linden, Paul J. 2008. "The ENSEMBLES Climate Change Project." In *Climate Variability, Modeling Tools, and Agricultural Decision-Making*, edited by Angel Utset, 65–106. New York: Nova.

WMO (World Meteorological Organization). 1992. *Global Aspects*. Vol. 1 of *Manual on the Global Data-Processing and Forecasting System*. Geneva, Switzerland: WMO.

———. 2002. "The WMO Statement on the Scientific Basis for, and Limitations of, Weather and Climate Forecasting." Abridged Report with Resolutions, 54th Session of the WMO Executive Council, Geneva, Switzerland. http://www.iabm.org/PDF_Files/Pages%20from%20941E-2.pdf.

———. 2003. "Socio-Economic Benefits of Meteorological and Hydrological Services." *WMO Bulletin* 52 (4): 366–73.

———. 2006. *Guidelines on the Role, Operation, and Management of National Hydrological Services*. Operational Hydrology Report 49. Geneva, Switzerland: WMO.

———. 2007. "Madrid Conference Statement and Action Plan: Secure and Sustainable Living—Social and Economic Benefits of Weather-, Climate- and Water-Related Information and Services." Madrid, March 19–22.

———. 2008. *WMO Guide to Meteorological Instruments and Methods of Observation*. Geneva, Switzerland: WMO.

———. 2009a. *Management of Water Resources and Application of Hydrological Practices*. Vol. 2 of *Guide to Hydrological Practices*. Geneva, Switzerland: WMO.

———. 2009b. "Position Paper on Global Framework for Climate Services." WMO, Geneva, Switzerland

———. 2010. *Implementation Plan for the Global Observing System for Climate in Support of the UNFCCC (2010 Update)*. Geneva, Switzerland: WMO.

———. 2011. *Regional Aspects*. Vol. 2 of *Manual on the Global Observing System*. Geneva, Switzerland: WMO.

———. 2012a. *Results of the Survey on the Role and Operation of National Meteorological and Hydrological Services Conducted in June–August 2011*. Geneva, Switzerland: WMO.

———. 2012b. "The WMO Strategy for Service Delivery." WMO, Geneva, Switzerland. http://www.wmo.int/pages/prog/amp/pwsp/documents/SDS.pdf.

Zillman, John W. 1999. "The National Meteorological Service." *WMO Bulletin* 48 (2): 129–59.

———. 2003a. "Meteorological and Hydrological Early Warning Systems." In *Early Warning Systems for Natural Disaster Reduction*, edited by Jochen Zschau and Andreas N. Kuppers, 135–64. Berlin: Springer.

———. 2003b. "The State of National Meteorological Services around the World." *WMO Bulletin* 52 (4): 360–65.

———. 2005. "The Challenges for Meteorology in the 21st Century." *WMO Bulletin* 54 (4): 224–29.

———. 2011. "Weather and Climate Information Delivery within National and International Frameworks." Paper presented at the First International Conference on Energy and Meteorology, "Weather and Climate for the Energy Industry," Gold Coast, Australia, November 8–11.

CHAPTER 4

Best Practices in Warning Systems

In This Chapter

Accurate information on meteorological and hydrological hazards—coupled with cross-government coordination with well-defined procedures for action—can result in effective responses that reduce loss of life and livelihoods.

Introduction

Early warnings of hydrometeorological hazards give people time to flee from a flash flood, tornado, or tsunami. They enable local authorities to evacuate or shelter large numbers of people in advance of a tropical cyclone or hurricane. They provide information on the occurrence of a public health hazard. And they enable a faster response to problems of food and water insecurity (World Bank 2010).

These warnings are vital because weather hazards and related events—such as hurricanes, heat waves, cold waves, windstorms, floods, and droughts—jointly cause more economic damage and loss of life than other natural disasters (GFDRR 2012; Gunasekera et al. 2004; UNISDR 2006b). Moreover, in recent decades such damage has been increasing, and climate change may make such events even more dangerous (IPCC 2007). Weather extremes also contribute to greater food insecurity, food price volatility, and poorer health.

In January 2005, at the World Conference on Disaster Reduction in Hyogo, Japan, 168 governments adopted a 10-year plan—the Hyogo Framework for Action (UNISDR 2005)—to make the world safer from natural hazards. The framework's goal is to substantially reduce disaster losses by 2015, not only of lives but also of the social, economic, and environmental assets of communities and countries. This global blueprint for disaster risk reduction efforts offers guiding principles and practical means for achieving disaster resilience for vulnerable communities. The framework identifies five priorities:

- Ensuring that disaster risk reduction is a national and local priority with a strong institution basis for implementation
- Identifying, assessing, and monitoring disaster risks and enhancing early warning

- Better managing knowledge for building a culture of safety
- Reducing the underlying risk factors
- Enhancing preparedness for an effective response

At this stage, several countries and regions have developed good practices in early warning and, in particular, in multihazard early-warning systems (MHEWSs). This chapter explores the role of National Meteorological and Hydrological Services (NMHSs) in these activities and the best practices that have been developed—such as the Shanghai MHEWS, one of the first of its kind, which has been used to train World Bank staff members and clients (World Bank and SMS 2012).

Effective Warning Systems

It should come as no surprise that the agencies best suited to issuing warnings of meteorological and hydrological hazards are the NMHSs, given their ability to identify, reduce, and transfer risks. They identify risks by (a) systematically observing and monitoring hydrometeorological parameters, (b) providing hazard forecasts and early warnings related to specific impacts to support emergency preparedness and response, and (c) providing historical and real-time hazard data to support financial risk transfer mechanisms. The encouraging news is that a systematic approach to managing the risks associated with disasters can prevent or mitigate their impact (see box 4.1 technical insight).

In developing a disaster risk management system, no single agency can provide a fully comprehensive solution. Therefore, it is vital that agencies work together, with a wide spectrum of stakeholders,[1] to narrow knowledge gaps and develop disaster risk management plans (WMO 2010; see box 4.2).

An effective warning system has three essential requirements:

- *Government leadership.* Such leadership supports policies and preparedness; supports organizing and coordinating disaster prevention and mitigation; and provides financial support for infrastructure and disaster relief.
- *Multiagency coordination.* Through coordination, agencies develop the warning platform and mechanisms that ensure intersectoral emergency response and interaction based on agreed levels of early-warning signals.
- *Community participation.* Communities participate by being prepared, holding drills, developing joint-preparedness teams, and raising awareness on self-rescue and mutual rescue.

The design and operation of severe weather warning systems must be based on a commitment to cooperate, exchange information, and form partnerships in the overall public interest (WMO 2010). In practice, building partnerships among government departments is challenging and difficult to accomplish without strong central government leadership, incentives, and a regulatory framework. Indeed, good governance is encouraged by robust legal and regulatory frameworks and is supported by long-term political commitment and effective

---

Box 4.1 Technical Insight: The Nature of Disaster Risks and Hazards

The concept of disaster risk or impact is used to describe the likelihood of harmful consequences arising from the interaction of natural hazards and the community. Three elements are essential in formulating disaster risk: (a) the probability of occurrence of a hazard, (b) the vulnerability of the community, and (c) the exposure of the community to that hazard. Thus,

$$\text{Disaster risk} = \text{Hazard probability} \times (\text{Exposure} \times \text{Vulnerability})$$

The three elements are defined as follows:

- *Hazard probability.* The hazard is the natural process of a phenomenon (flood, storm, drought) with adverse effects on life, livelihoods, and property. By seeking to understand hazards of the past, monitoring the present, and predicting the future, a community or public authority is poised to minimize the risk of a disaster. NMHSs play a key role in this aspect of risk management of weather-related natural disasters by determining the hazard probability.
- *Exposure.* The people and property subject to the hazard are called the *exposure*.
- *Vulnerability.* This term refers to a characteristic that influences damage. The community's vulnerability is the susceptibility and resilience of the community and the environment to natural hazards. Different population segments can be exposed to greater relative risks because of their social and economic conditions. Reducing disaster vulnerability (mitigation) requires (a) increasing knowledge about the likelihood, consequences, imminence, and presence of natural hazards and (b) empowering individuals, communities, and public authorities with that knowledge to lower the risk before severe weather events occur (preparedness) and to respond effectively immediately afterward (relief).

---

institutional arrangements. Effective governance arrangements should encourage local decision making and participation, supported by broader administrative and resource capabilities at the national or regional level. Vertical and horizontal communication and coordination among early-warning stakeholders are also essential.

One important factor in early warning is the trust people have in the institution responsible for the warning. Experience shows that this trust is very difficult to create, but it is easy to destroy. In practical terms, it is preferable to be conservative in what products are distributed to the public; it is better to wait to make sure that a product is appropriate and useful, rather than making public a half-baked product that will undermine confidence in the institution. And the NMHSs need to have direct contact with the population and some visibility to ensure that this trust is created.

NMHSs must understand the decision-making processes of all the sectors affected by the hazard to ensure that information is tailored to the users' specific needs. Tailoring information to specific users involves efficient and timely synthesis and elucidation of weather-related data and information and their effect on

> **Box 4.2 Benefits of Partnerships for Early-Warning Systems**
>
> The benefits of partnerships for early-warning systems are many:
>
> - Sharing meteorological and hydrological expertise for flood warning[a] and seismological and oceanographic expertise for tsunami warning[b]
> - Drawing expertise from a wide range of disciplines, such as social science, community planning, and engineering
> - Accomplishing tasks that cannot be managed by a single agency or organization
> - Demonstrating to government budget planners a commitment to work together toward a common goal and making better use of scarce financial resources
> - Leveraging resources for research, awareness, and preparedness
> - Sharing costs, knowledge, and lessons learned
> - Ensuring a consistent message (the warning bulletins and other outreach material) from multiple credible sources
> - Distributing messages more widely through multiple outlets and receiving feedback from a wide range of users
> - Optimizing the effectiveness of warnings with regard to fewer fatalities and avoiding unnecessary recovery costs
>
> a. The Met Office and the Environment Agency in the United Kingdom operate a Joint Flood Forecasting Center.
> b. The Bureau of Meteorology and Geoscience Australia operate a Joint Tsunami Warning Center.

the users' operations and objectives. It also includes a quantitative understanding of the social and economic cost and benefit of warnings (Rogers and Tsirkunov 2011). Maximizing the benefit depends on understanding the uncertainty in the warning, the decisions that depend on the warning, and the level of acceptance of false alarms.

A warning system must empower individuals, communities, and businesses to respond to hazards in a timely and appropriate manner that will reduce the risk of death, injury, property loss, and damage (see box 4.3). Precise and longer lead times of warnings are increasingly needed, using probabilistic forecasts. To be effective, early-warning messages must be understood. The must provide the right information to the right people at the right time and right place—what is commonly referred to as a *people-centered early-warning system*.

Core Elements of a Warning System

What are the core elements of a people-centered system? They range from planning knowledge of the risks faced to preparedness to act—and failure in any one of these elements might mean failure of the whole system (figure 4.1):

- *Risk knowledge.* Disaster risks arise from the combination of hazards, vulnerabilities, and exposure at a particular location. Identifying and assessing risk require the systematic collection and analysis of data. Such a risk assessment

Box 4.3 Ranking the Priority of Weather Warning Services

Which warning services are the most prized? According to the World Meteorological Organization's 2000–01 global survey, priority weather warning services are ranked as follows: (a) severe thunderstorms, (b) tropical cyclones, (c) land gales, (d) hail, (e) blizzards, (f) fire weather, (g) midlatitude cyclones, (h) sandstorms and dust storms, and (i) tornadoes (ranked much higher in individual countries than for the world as a whole).

As for hydrological hazards, the global ranking is (a) flash floods, (b) river floods, (c) erosion and landslides, and (d) algal outbreaks.

Figure 4.1 Elements of a People-Centered Early-Warning System

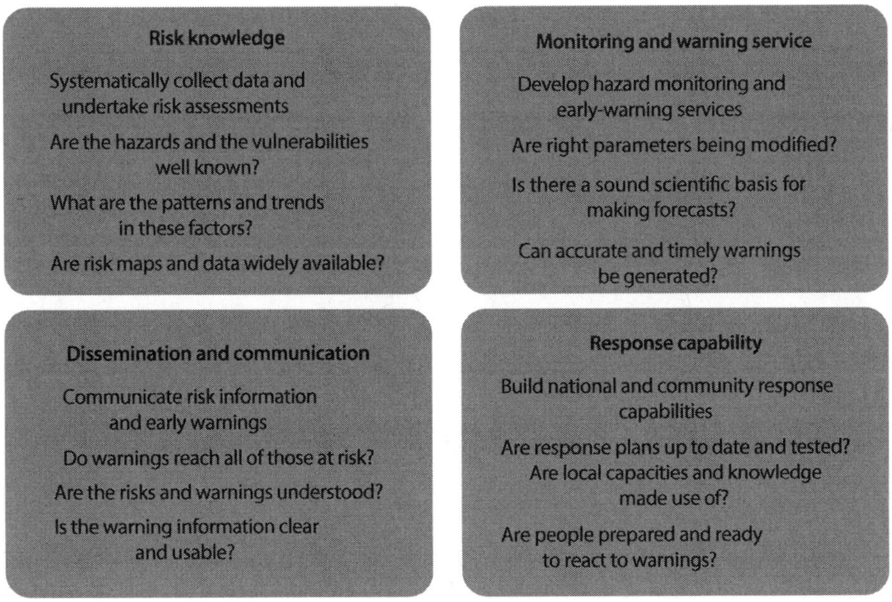

Risk knowledge	Monitoring and warning service
Systematically collect data and undertake risk assessments	Develop hazard monitoring and early-warning services
Are the hazards and the vulnerabilities well known?	Are right parameters being modified?
What are the patterns and trends in these factors?	Is there a sound scientific basis for making forecasts?
Are risk maps and data widely available?	Can accurate and timely warnings be generated?
Dissemination and communication	**Response capability**
Communicate risk information and early warnings	Build national and community response capabilities
Do warnings reach all of those at risk?	Are response plans up to date and tested?
Are the risks and warnings understood?	Are local capacities and knowledge made use of?
Is the warning information clear and usable?	Are people prepared and ready to react to warnings?

Source: UNISDR 2006a.

should weigh the dynamic nature of hazards, vulnerabilities, and exposure that arise from processes such as urbanization, rural land-use change, environmental degradation, and climate change. Risk assessments and maps help motivate people, prioritize early-warning system needs, and guide preparations for disaster prevention and responses.

- *Monitoring and warning service.* There must be a sound scientific basis for predicting and forecasting hazards, along with reliable forecasting and warning systems that operate 24 hours a day.[2] Continuous monitoring of hazard parameters and precursors is essential to generate accurate warnings in a timely fashion. Warning services for different hazards should be coordinated where possible to gain the benefit of shared institutional, procedural, and communication networks. This benefit can be achieved through a MHEWS that coordinates and integrates the needs of different stakeholders. Traditionally, NMHSs have focused on providing warning information directly linked to the hydrometeorological forecasts that they calculate. But the impact of heavy precipitation, for example, will vary over a catchment, with some people being at little risk while others are in life-threatening situations. Thus, targeting those at risk creates a more effective response and reduces the risk of warning fatigue and false alarms.

- *Dissemination and communication.* Clear messages containing simple, useful information are critical to enable proper responses that will help safeguard lives and livelihoods. Regional, national, and community-level communication systems must be preidentified and appropriate authoritative voices established. The use of multiple communication channels is necessary to ensure that as many people as possible are warned, to prevent the effects of failure of any one channel, and to reinforce the warning message.

- *Response capability.* Education and preparedness programs play a key role in ensuring that communities understand their risks, respect the warning service, and know how to react. It is also essential that disaster management plans be in place, well practiced, and tested. The community should be well informed on the options for safe behavior, the available escape routes, and the best ways to prevent damage and loss to property.

In addition, people-centered early-warning systems need to take into account a number of cross-cutting issues, such as involving local communities; considering gender perspectives, cultural diversity, and disability issues; and taking a multihazard approach. A system that serves the needs of numerous agencies should provide significant cost savings over systems that would otherwise be developed separately for each agency (see box 4.4).

How Multihazard Warning Systems Work

The World Meteorological Organization (WMO) has identified seven national good practices and guiding principles in MHEWSs (Golnaraghi 2012). One example that highlights a complete end-to-end system, which in principle could be adapted to other countries, is the Shanghai Meteorological Service (SMS) MHEWS. This system is set up to provide warnings of hydrometeorological hazards in a megacity with more than 23 million people, where much of the infrastructure is vulnerable to disruption (Tang et al. 2012). The Shanghai MHEWS is comprehensive. It is designed to cope with the threats from tropical cyclones, storm surges, rainstorms, heat and cold waves, and thunderstorms, as well as the cascading threats that they cause, such as floods, health impacts, accidents, and infrastructure damage (see photo 4.1).

Box 4.4 Cross-Cutting Issues for People-Centered Early-Warning Systems

Involving Local Communities

People-centered early-warning systems rely on the direct participation of those most likely to be exposed to hazards. Without the involvement of local authorities and communities at risk, government and institutional interventions and responses to hazard events are likely to be inadequate. A local bottom-up approach to early warning, with the active participation of local communities, enables a multidimensional response to problems and needs. In this way, local communities, civic groups, and traditional structures can contribute to reducing vulnerability and strengthening local capacities.

Considering Gender Perspectives, Cultural Diversity, and Disability Issues

It is essential to recognize that different groups have dissimilar vulnerabilities, according to culture, gender, or other characteristics that influence their capacity to effectively prepare for, prevent, and respond to disasters. Women and men often play different roles in society and have different access to information in disaster situations. In addition, elderly people, people with disabilities, and socioeconomically disadvantaged people are often more vulnerable. Information, institutional arrangements, and warning communication systems should be tailored to meet the needs of every group in every vulnerable community.

Taking a Multihazard Approach

Where possible, early-warning systems should link all hazard-based systems. Economies of scale, sustainability, and efficiency can be enhanced if systems and operational activities are established and maintained within a multipurpose framework that considers all hazards and end users' needs. Multihazard early-warning systems (MHEWS) will also be activated more often than a single-hazard warning system and, thus, should provide better functionality and reliability for dangerous high-intensity events (such as tsunamis) that occur infrequently. And multihazard systems help the public better understand the range of risks and reinforce desired preparedness actions and warning response behaviors.

The Shanghai MHEWS is based on two core concepts:

- Establishing laws, regulations, and standardized operating procedures and mechanisms for a multiagency response—which clearly identify roles and responsibilities
- Providing operating procedures for early detection, briefing, and warning dissemination on the basis of good observations and forecasts

Legal Framework

The legal process in China began in 2006 when national regulations were implemented for prevention and emergency response preparedness and monitoring and warning. In addition, the Meteorology Law of China defines the roles, responsibilities, and authorities of weather departments, as well as their

Photo 4.1 Part of the Shanghai Meteorological Service's delivery platform for public weather services and multi-hazard warnings.

operational functions. And the Flood-Control Law of China also includes requirements for meteorological services to provide weather forecasts to flood-control headquarters.

At the city level, the Shanghai government passed regulations implementing the Meteorology Law of China and clarifying the role of the SMS in disaster prevention and mitigation—a framework that continues to be strengthened. At least 15 separate ministries and departments must coordinate with the municipal government, each of which has signed agreements that establish operating procedures.

Operating Procedures

Given that the multihazard system focuses on managing the potential cascade of disasters stemming from an initial hydrometeorological hazard, the primary, secondary, and sometimes tertiary impacts require well-ordered coordination and cooperation to support highly sensitive users as well as the general public. This multiagency coordination and multiphase response requires standard operating procedures (SOPs) and has led to the concept of the *five earlys*: (a) early monitoring and warning, (b) early briefing (for special users and agencies well in advance of public warnings), (c) early warning, (d) early dissemination, and (e) early handling.

The MHEWS consists of two components:

- The *management component* involves (a) a multiagency coordination and cooperation mechanism consisting of government organizations and (b) a social

community protection system consisting of the basic social units, such as communities.
- The *technical component* has six platforms: (a) multihazard detection and monitoring, (b) forecast and warning information generation, (c) decision-making support, (d) warning information dissemination, (e) a multihazard database, and (f) a multiagency network system.

The platforms for detection, monitoring, and forecasting are extensive, including a dense network of in situ and remotely sensed observations and numerical weather prediction models (figure 4.2). These models provide the forecaster with the tools to support the decision-making platform, which includes all aspects of information sharing and communication as part of the SMS's public weather service function.

The warning information dissemination platform ensures that messages are consistent and actionable, disseminated through multiple channels, communicated quickly, and received effectively. As needed, at this stage, weather- and

Figure 4.2 Operational Flow of the Shanghai Meteorological Service Forecasting and Warning System

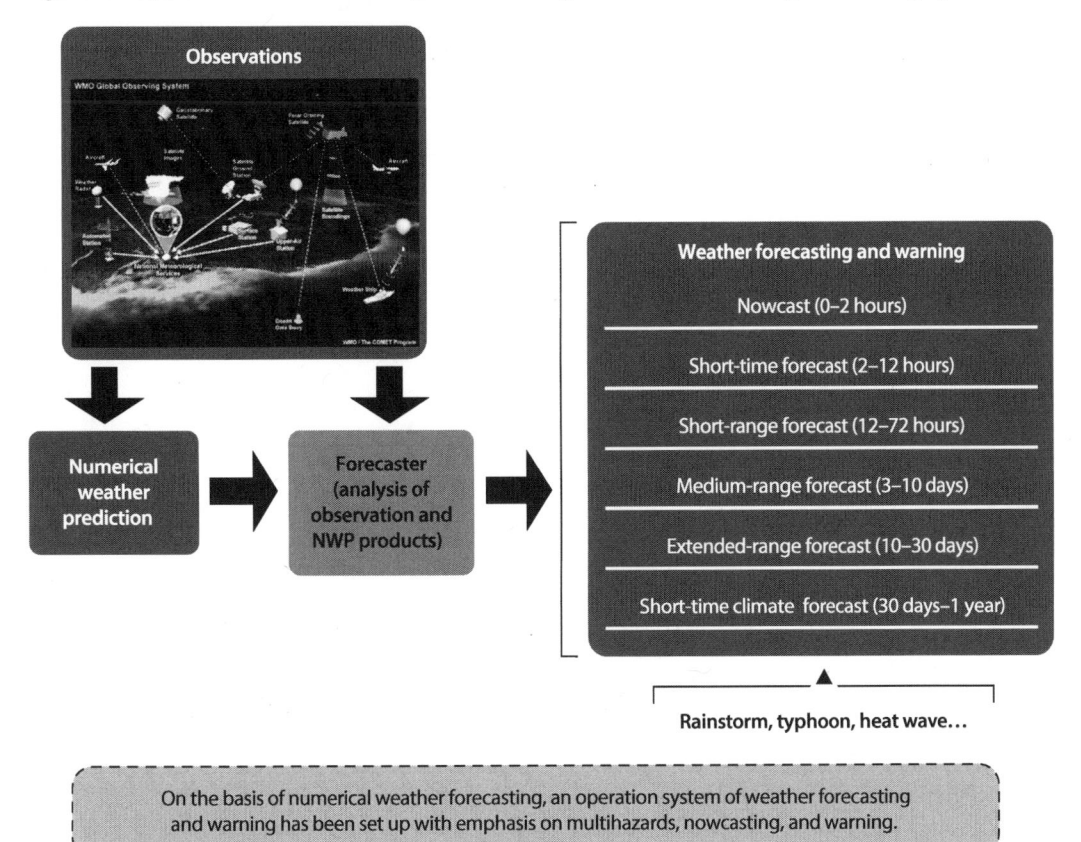

On the basis of numerical weather forecasting, an operation system of weather forecasting and warning has been set up with emphasis on multihazards, nowcasting, and warning.

Source: Tang Xu, Shanghai Meteorological Service.
Note: NWP = numerical weather prediction.

Figure 4.3 Severe Weather Warning Signals in Shanghai

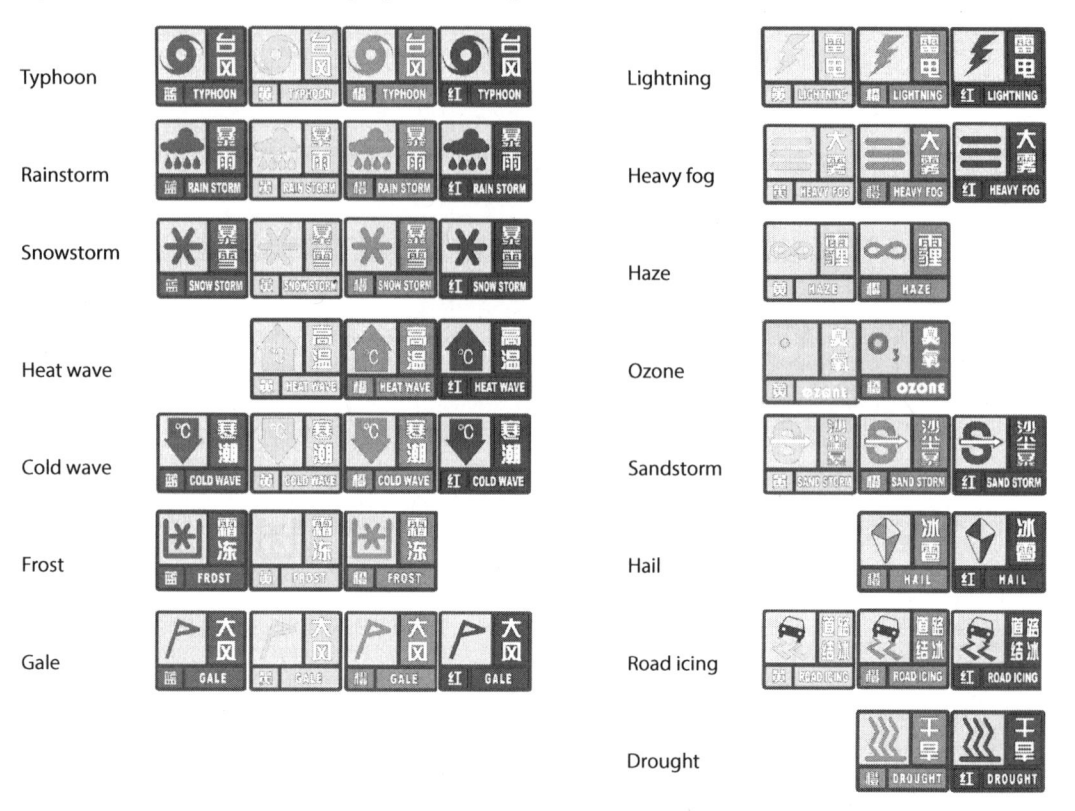

Typhoon		Lightning
Rainstorm		Heavy fog
Snowstorm		Haze
Heat wave		Ozone
Cold wave		Sandstorm
Frost		Hail
Gale		Road icing
		Drought

Source: Tang Xu, Shanghai Meteorological Service.
Note: There are a total of 15 categories and 46 warnings in this system.

climate-related hazard information is combined with other hazard information. The dissemination system uses color-coded symbols, which are consistent across all agencies responsible for providing warnings. They are disseminated by television, telephone, websites, warning towers, radio, mobile short messages, and large electronic displays (figure 4.3).

The database platform consists of historical and real-time data supplied by all the agencies and sectors at risk, including water affairs, maritime, rural traffic, air traffic, food, sanitation, agriculture, and electric power. More than 17 departments currently support this database.

The platform for the multiagency network system supports the development of SOPs. Currently, there are 36 joint-response mechanisms among 25 government departments. This approach facilitates efficient cooperation in emergency management. Early briefing prepares the departments to act ahead of the joint-response mechanisms and before warnings are issued to the public.

Another important element in ensuring the system's efficiency is incorporating the warning system into the public weather service[3] operations. This system is anchored by the public service officer, who is responsible for disseminating both routine and hazard-related information (figure 4.4).

Figure 4.4 Public Weather Service Work Flow

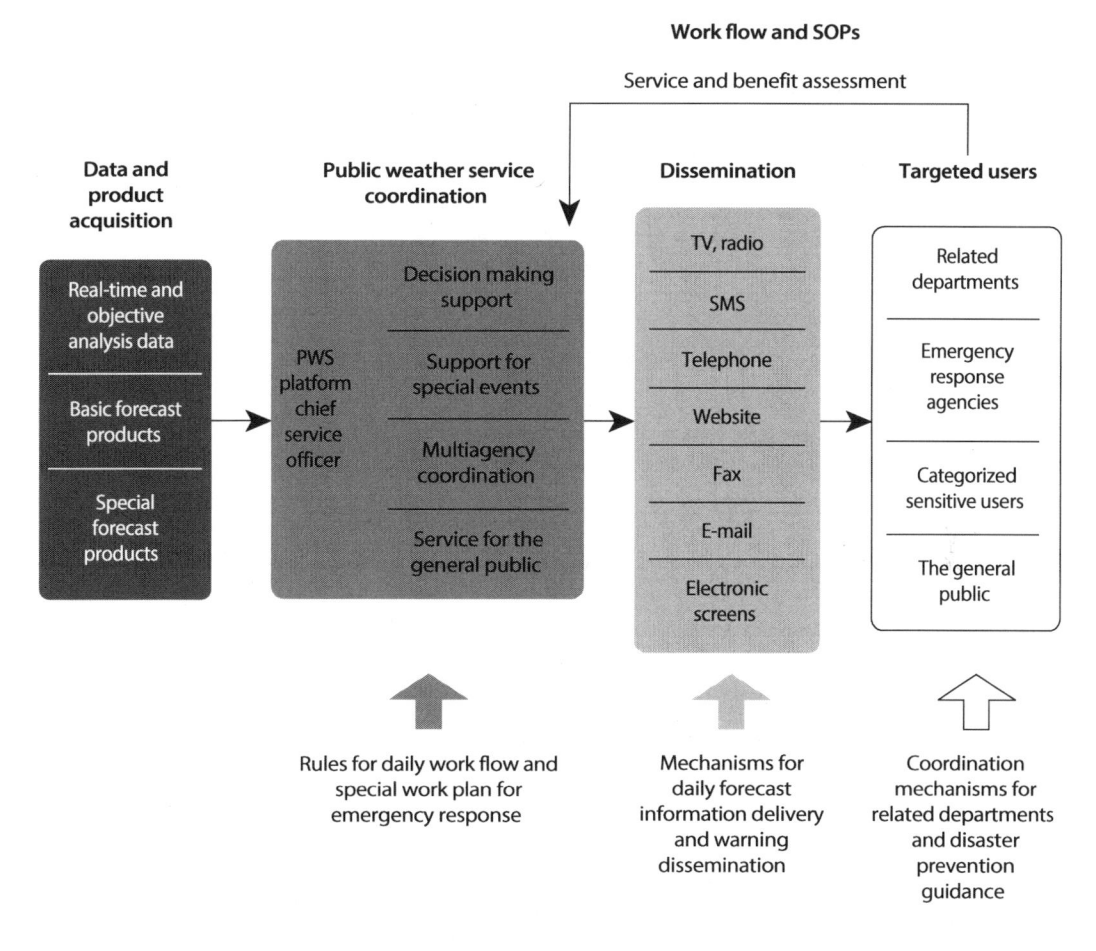

```
                                    Work flow and SOPs
                                Service and benefit assessment

   Data and          Public weather service      Dissemination        Targeted users
   product              coordination
  acquisition
                                                   TV, radio           Related
                       Decision making                                departments
 Real-time and            support                    SMS
  objective
 analysis data   PWS                               Telephone          Emergency
                 platform  Support for                                 response
                 chief    special events                               agencies
 Basic forecast  service                            Website
   products      officer  Multiagency
                          coordination                Fax             Categorized
                                                                    sensitive users
   Special                Service for the           E-mail
  forecast                general public                              The general
  products                                         Electronic           public
                                                    screens

              ⬆                              ⬆                      ⬆
       Rules for daily work flow and    Mechanisms for         Coordination
         special work plan for          daily forecast         mechanisms for
          emergency response         information delivery   related departments
                                         and warning         and disaster
                                         dissemination         prevention
                                                                guidance
```

Source: Tang Xu, Shanghai Meteorological Service.
Note: CSO = chief service officer; PWS = public weather service; SMS = short message service; SOP = standard operating procedure.

Lessons from Shanghai's Multihazard System

What lessons can countries learn from Shanghai's experience with the MHEWS? The World Bank has recently partnered with the China Meteorological Administration and the SMS to share China's experience in developing early-warning systems. The first of a series of workshops, which was held in Shanghai in March 2012, suggested the following broad lessons (World Bank and SMS 2012):

- *A meteorological or hydrometeorological service should evolve.* The SMS has transformed from a traditional weather agency to a user-oriented service organization focused on delivering services that people need and want. This change is achieved by creating a dialogue with the public, as well as specific weather- and climate-sensitive sectors and government agencies.

Although the technological advances are important, it is recognized that they must follow the identification of needs and human capacity. Communication comes first, then the technology. Different communication skills are needed to enable different users to understand the warning process. Finding commonality with target groups is essential.

- *The emphasis should be on delivering services.* The Shanghai experience confirms the World Bank's approach, which is to go beyond the technical level and ensure that the overall goal is service delivery. Weather- and climate-related warning information should be delivered from a single platform. Knowing how to work with other agencies is essential. For developing countries, the challenge is how to achieve this delivery scheme with limited capacity and limited financial resources. One approach is to focus on developing a physical center of excellence within the National Meteorological Service (NMS) or national hydrological service and use it to demonstrate how a multihazard system can be created—with the emphasis on how to avoid problems. Every effort should be made to apply the expertise and experience of exemplars (such as the SMS and the China Meteorological Administration) in institutional and financial environments that are less conducive. The Shanghai MHEWS concept can be applied with necessary adjustments in least developed countries.

- *The government needs to have a strong political commitment.* This political commitment is the foundation of developing a coordinated, multiagency, MHEWS. A legal and regulatory framework is vital so that responsibilities are clearly differentiated and SOPs can be achieved. The political commitment establishes an expectation that agencies will cooperate with NMHSs—and this expectation creates a strong foundation for collaboration.

- *Training should be expanded to include users of an NMS's services.* Programs need to be developed with advanced NMSs and the WMO that target the specific requirements of World Bank clients. This objective is especially important because warning services extend from the primary weather warnings for civil protection to other affected areas, such as health, food, and water security.

- *Standards and best practices are essential.* The WMO stresses the importance of standards and best practices, but currently no common worldwide practice exists. The Shanghai MHEWS approach focuses on risk reduction systems that can be applied bottom up within developing countries. Applying this experience may encourage sharing data, information, and know-how among developing countries as a form of South-South cooperation. It has been suggested that the World Bank could play a role in developing and implementing standards by helping countries apply relevant SOPs (such as warnings, technology, and information). Other examples of best practices include the multihazard warning system of the U.S. National Weather Service (NWS; Keeney, Buan, and Diamond 2012), the vigilance system operated by Météo France (Borretti and Degrace 2012), and the MHEWS of Japan (Hasegawa et al. 2012).

In addition, given that many World Bank clients will be unable to completely develop stand-alone forecasting services and that such services might not even be desirable, it would make sense for them to make greater use of the WMO's Regional Specialized Meteorological Centers (RSMCs). In fact, strengthening and using RSMCs to help support national services would be a cost-effective way of helping countries develop and sustain their services. Use of these centers could strengthen regional numerical weather prediction and help countries customize their services, thus creating a unified warning system based on the increasingly common practice of the four-color system (green, yellow, orange, and red representing different levels of alert) used in China and many European countries. Use of RSMCs would also enable countries to focus more on the principal role of communicating warnings and delivering other services to the public and weather- and climate-sensitive sectors.

Collaboration among regions and centers is vital for sharing expertise and improving information services—such as the Severe Weather Information Centre in Hong Kong SAR, China, and the World Weather Information Service—but not enough members are currently contributing information to these websites. More consistency between regional centers is also needed. Different RSMCs issue tropical cyclone warnings, but formats are inconsistent, making it difficult to create a fully integrated system. Another issue is arranging sustainable financing for the RSMCs. It has been recommended that the World Bank work with the WMO on identifying and testing long-term financial instruments to make the RSMCs' performance more sustainable and effective.

In the end, effective warning systems depend on the capacity of different government departments at all levels—national, provincial, and local—to cooperate. Such cooperation is exemplified by Hurricane Sandy, which caused approximately US$30–US$50 billion in damages and a loss of more than 110 lives in the United States alone. Right after Sandy hit, expert analyses lauded U.S. readiness and response (see box 4.5). But further analysis has revealed some weaknesses in the system:

- The U.S. NWS forecast and warning caused confusion as the storm moved into the New Jersey–New York area. The NWS changed the category of the storm from a hurricane to an extratropical cyclone because it no longer met the technical criteria for a hurricane or tropical storm. This change had little to do with the storm's intensity, which did not change significantly. Nevertheless, the public perceived a weaker storm, and in some areas, people let their guard down.
- Overall, the state and local governments and the public did not appear to understand the impending impact of the storm surge and coastal flooding.
- Recovery has been slow.

Box 4.5 Drawing Lessons from Hurricane Sandy

The following is an excerpt from the *Living on the Real World* blog (Hooke 2012):

> Viewed narrowly Hurricane Sandy is a success story. Start with the forecast. Americans were given a week's heads-up that Hurricane Sandy would track north and then instead of veering safely out to the Atlantic, would come ashore somewhere near New Jersey and slowly work inland before reorganizing and heading north through Canada....
>
> Then there's the emergency response. Emergency managers took fullest advantage of their week to prepare. We saw a remarkable mobilization at federal, state, and local levels, accompanied by private-sector collaboration with respect to critical infrastructure: the power grid, communications, gas and water utilities, sewage, and much more. There was some roughness around the edges. The normal emergency procedures were overwhelmed by the severity of events at a number of points. There was some political-level friction across state boundaries and between state and local levels. But still and all, the response maintained remarkable focus, combining with media coverage to keep the U.S. death toll [low]....
>
> Add it all up? America is growing more skilled—and getting better fast—at emergency response to disasters of growing geographical reach, cost, and complexity.
>
> But we can and should do more. [More than 110] lives lost to Sandy, [on top of] ... the seventy deaths reported from the Caribbean, ... represents too much grief and suffering. That early estimate of $10–$20B in losses has already escalated to $5–$10B insured losses and overall costs of $30–$50B. Any final accounting will probably show the cost of this disaster to be more comparable to Hurricane Katrina than Hurricane Irene. A big hit even for the U.S. $14T economy just as it's finally starting to recover from the financial-sector meltdown of 2008. The prospect of a continuing stream of such events in the future of ever-greater magnitude? Unacceptable....
>
> America needs a comparable national effort and accompanying long-term investment in reducing the need for emergency response on such a grand scale.
>
> The need for emergency response will never go away. But we shouldn't resign ourselves to the idea that emergencies will necessarily continue to grow in scope, number, and impact, just because our society is growing in numbers, in property exposure, and in economic activity. We can grow our society's resilience to such events. We can reduce the geographical extent and the population adversely affected by future events.
>
> We actually have a shining example, one we can build on:
>
> Commercial aviation.... Over the past fifty years, property loss to natural hazards has been growing exponentially ... [partly] the result of growing population and property exposure in hazardous areas. But it's also the result of a failure to learn from experience; an insistence on "rebuilding as before." By contrast, commercial air travel as measured by takeoffs and landings has quadrupled over the same half-century, but the

box continues next page

Box 4.5 Drawing Lessons from Hurricane Sandy (continued)

number of flight-related accidents has remained constant over that period or even declined. That's because of a remarkable public-private partnership on the part of the airlines and the [Federal Aviation Administration], and because of the catalytic role played by a small but vital independent federal agency, the National Transportation Safety Board (NTSB). The NTSB mantra is not "the wing fell off this airplane, but we're going to rebuild it as before," but rather "What caused this accident? We have to make sure it never happens again."

We need an analog to the NTSB for natural hazards. Each catastrophe should trigger a national conversation, not just at the federal but also state and local levels along the lines of "what can our community here learn from what happened (over there)?" And that conversation should lead to a set of mutually supportive private- and public-sector actions to build resilience at the community level and reduce future risk.

Overall, it is clear that maintaining an effective warning and response system anywhere is a continuous and ongoing process that needs to be frequently tested, evaluated, and improved by NMHSs, emergency managers, and communities at risk. Important enabling factors are the way NMHSs operate and how they are financed and regulated. We turn to this issue in the next chapter.

Notes

1. Stakeholders include government agencies with missions involving the protection of life and property; national, regional, or local emergency management agencies; first responders; and infrastructure managers (dams, transportation departments, and bridges). Other stakeholders include the media; nongovernmental organizations; emergency relief and humanitarian organizations, such as the International Red Cross and the Red Crescent Society; academic institutions and schools; trained volunteers associated with National Meteorological and Hydrological Services, such as cooperative observers, storm spotters, and amateur radio operators; meteorological societies and other professional associations in risk management disciplines; private sector weather companies; and utility services, telecommunication operators, and other operation-critical or weather-sensitive businesses.

2. The accuracy of warnings of high-impact weather is improved by the routine day-to-day forecasting operations of a National Meteorological Weather Service (NMWS). Forecasting skills remain high when exercised frequently and when training related to specific extreme events is part of the routine activities of forecasters. For example, the U.S. NWS ensures that forecasters are well prepared for the hurricane and severe convective seasons by through its National Hurricane Center and Storm Prediction Center.

3. For more information on the function of public weather services, visit World Meteorological Organization's website at http://www.wmo.int/pages/prog/amp/pwsp/wmopws_en.htm.

References

Borretti, Catherine, and Jean-Noel Degrace. 2012. "The French Vigilance System: Contributing to the Reduction of Disaster Risks in France." In *Institutional Partnerships in Multi-Hazard Early Warning Systems*, edited by Maryam Golnaraghi, 63–93. Heidelberg, Germany: Springer.

GFDRR (Global Facility for Disaster Reduction and Recovery). 2012. *The Sendai Report: Managing Disaster Risks for a Resilient Future*. World Bank, Washington, DC.

Golnaraghi, Maryam, ed. 2012. *Institutional Partnerships in Multi-Hazard Early Warning Systems*. Heidelberg, Germany: Springer.

Gunasekera, Don, Neil Plummer, Tony Bannister, and Linda Anderson-Berry. 2004. "Natural Disaster Mitigation: Role and Value of Warnings." Paper presented at the Australian Bureau of Agricultural and Resource Economics 2004 Outlook Conference, Disaster Management Workshop, Canberra, March 2–3.

Hasegawa, Naoyuki, Satoshi Harada, Shotaro Tanaka, Satoshi Ogawa, Atsushi Goto, Yutaka Sasagawa, and Norihisa Washitake. 2012. "Multi-Hazard Early Warning System in Japan." In *Institutional Partnerships in Multi-Hazard Early Warning Systems*, edited by Maryam Golnaraghi, 181–216. Heidelberg, Germany: Springer.

Hooke, William. 2012. "Hurricane Sandy's Real Lesson ... Will We Learn It?" *Living on the Real World* (blog), October 31. http://www.livingontherealworld.org/?p=755.

IPCC (Intergovernmental Panel on Climate Change). 2007. *Climate Change 2007: The Physical Science Basis—Contribution of Working Group I to the Fourth Assessment Report of the Intergovernmental Panel on Climate Change*. Cambridge, U.K.: Cambridge University Press.

Keeney, Harold Jr., Steve Buan, and Laura Diamond. 2012. "Multi-Hazard Early Warning of the United States National Weather Service." In *Institutional Partnerships in Multi-Hazard Early Warning Systems*, edited by Maryam Golnaraghi, 115–57. Heidelberg, Germany: Springer.

Rogers, David P., and Vladimir Tsirkunov. 2011. "Costs and Benefits of Early Warning Systems." In *Global Assessment Report on Disaster Risk Reduction*. Geneva, Switzerland: United Nations.

Tang, Xu, Lei Feng, Yongjie Zou, and Haizhen Mu. 2012. "The Shanghai Multi-Hazard Warning System: Addressing the Challenge of Disaster Risk Reduction in an Urban Megalopolis." In *Institutional Partnerships in Multi-Hazard Early Warning Systems*, edited by Maryam Golnaraghi, 159–79. Heidelberg, Germany: Springer.

UNISDR (United Nations International Strategy for Disaster Reduction). 2005. "Hyogo Framework for Action (HFA): Building the Resilience of Nations and Communities to Disasters." UNISDR, Geneva, Switzerland. http://www.unisdr.org/eng/hfa/hfa.htm.

———. 2006a. "Developing Early Warning Systems: A Checklist." UNISDR, Geneva, Switzerland. http://www.unisdr.org/2006/ppew/info-resources/ewc3/checklist/English.pdf.

———. 2006b. "Disaster Statistics, 1991–2005." UNISDR, Geneva, Switzerland. http://www.unisdr.org/we/inform/disaster-statistics.

WMO (World Meteorological Organization). 2010. *Guidelines on Early Warning Systems and Application of Nowcasting and Warning Operations*. Geneva: WMO. http://www.wmo.int/pages/prog/amp/pwsp/documents/PWS-21.pdf.

World Bank. 2010. *Natural Hazards, Unnatural Disasters: Effective Prevention through an Economic Lens*. Washington, DC: World Bank.

World Bank and SMS (Shanghai Meteorological Service). 2012. *Multi-Hazard Early Warning Decision Support Systems Workshop Report*. Washington, DC: World Bank; Shanghai, China: SMS. http://www.gfdrr.org/gfdrr/sites/gfdrr.org/files/Multi-Hazard_Early_Warning_and_Decision_Support_Systems_Workshop_12-04-13.pdf.

CHAPTER 5

Financing, Operating Models, and Regulatory Frameworks

In This Chapter

No single preferred operating model exists for either National Meteorological Services (NMSs) or National Hydrological Services (NHSs). But given the primary mission of the National Meteorological and Hydrological Services (NMHSs) to save lives and property, governments need to retain significant responsibility for the services' operations and ensure that this duty is adequately resourced. If too much autonomy is granted, NMHSs can be pushed into a competitive situation with the private sector and other government agencies, which can weaken their capacity to deliver mandated services. An unambiguous legal and regulatory framework is also essential.

Introduction

To mitigate weather-, climate-, and water-related risks, NMHSs must have the means to sustain and ensure the ongoing relevance of their mandated services. The inclusion of new capabilities, the creation of new services, and operations and maintenance require additional and ongoing budget support. Although these costs may be covered by greater core support, such funds are not always available. Therefore, carefully developed financing and management plans are needed.

Many governments have sought ways to reduce the cost of providing meteorological and hydrological services as a part of an overall strategy to reduce public spending. Common approaches have been to transfer NMHSs from government departments to more independent government agencies or, rarely, to outsource the provision of weather services to the private sector. As a result, the obligation often falls on the NMHSs to increase revenues to meet the additional costs.

So how should a government choose an effective operating model for its national meteorological or hydrometeorological service[1] that is tailored to the country's specific needs? Unfortunately, despite the importance of such models, there is little documented experience in developing and implementing them for

NMHSs. This chapter looks at the public sector models—ranging from no autonomy to a great deal of autonomy—that are currently used by NMHSs globally. It also explores the best types of unambiguous legal and regulatory frameworks.

Organization of NMHSs

NMHSs operate under a range of titles, from Department of Meteorology and Hydrology to National Weather Service and may have a service focus, a research focus, or both. Countries may have separate meteorological and hydrological services. But unlike operational meteorology, which is normally within a single national institution, operational hydrology and water resource assessment activities may be shared among several government departments.[2] There are about 80 stand-alone NHSs with considerable diversity among them—for example, the River Bureau in Japan, the Ministry of Water and Irrigation in Kenya, and the Geological Survey in the United States.[3]

Typically, the closer these activities are aligned, the more effective the warning services for flood-related hazards are, given that flood forecasts require both meteorological and hydrological inputs. In most countries, the meteorological and hydrological warning services are integrated fully into the central operations of the NMHSs. In some countries, however, separate warning centers exist within the individual NMSs or regional service offices.

Governments are beginning to encourage joint warning centers, such as the Flood Forecasting Centre in the United Kingdom, which is operated by the Met Office (the United Kingdom's NMS) and the Environment Agency,[4] and the Shanghai Multihazard Early-Warning System, which is operated by the Shanghai Meteorological Service on behalf of the Shanghai municipal government. The United Kingdom has also created a natural hazards partnership. Under this partnership, the Met Office, the Environment Agency, the Health Protection Agency, the Cabinet Office, and currently nine other agencies provide a single cohesive picture of the potential impact of environmental hazards by using predictive tools, made possible by pooling data and information from each of the partners (Gray 2012).

Establishing a legal framework for forecasts and warning services is essential to ensure that all stakeholders understand their roles and responsibilities. The organizational structures through which NMSs perform their functions and deliver their services vary widely (see box 5.1).

Funding

For most of the 20th century, nearly all governments accepted full responsibility for funding the operation of their NMHSs, which have been viewed along with the police, army, national radio broadcaster, and a few other entities as an essential component of the national taxpayer-funded government infrastructure. This view has provided a sound international framework for conducting activities whose benefits apply widely to the community and to future generations, but it has not made it easy to assemble the necessary resources for effective operation

Box 5.1 Organization and Staffing of National Meteorological Services

Situation of National Meteorological within Governments

A survey conducted by Environment Canada found that 33 percent of the National Meteorological Services (NMSs) were placed in departments of environment, 30 percent were separate agencies, 23 percent were in departments of transport, 3 percent were in departments of defense, and 10 percent were in departments of agriculture (Jean et al. 1999).

The latest World Meteorological Organization (WMO) survey shows that about 27 percent of NMSs are located within ministries of environment, 22 percent in ministries of transport, 9 percent in ministries of science, 5 percent in ministries of agriculture, 5 percent in ministries of defense, 5 percent in ministries of communication, and 3 percent in ministries of education (WMO 2012). About 23 percent of the respondents were not associated with any of the seven specified ministries.

Staffing

The total staffing of NMSs worldwide is about 300,000, with the largest exceeding 50,000 employees and the smallest having fewer than 5 (Zillman 2003). In 2011, the highest staffing levels reported were in China (53,599), the Russian Federation (about 36,000), Indonesia (4,413), France (3,614), and Kazakhstan (3,132). The lowest staffing levels were reported by Malta (14), Monaco (17), Tonga (28), Antigua and Barbuda (31), Benin (31), St. Lucia (31), Niger (45), Cameroon (54), and Costa Rica (89).[a] On average, the makeup of NMS staff is around 45 percent professional, 45 percent technical, and 10 percent administrative. Meteorological employees per 1,000 square kilometers range from fewer than 1–20 or more (Zillman 1999, 2003).

Education Levels

About half of all staff members have a high school education, 30 percent have a bachelor's degree, and 20 percent have an advanced degree. In developed and transitional economy countries, the percentages of employees with degrees are significantly higher than in developing and least developed countries. The number of NMS directors who have held their post for fewer than five years has increased from 50 percent (according to the 2000–01 WMO survey) to 71 percent, according to the latest survey (WMO 2012).

Indicative of the importance of highly qualified personnel, the China Meteorological Administration increased the number of staff members with bachelor's or advanced degrees from 10,953 to 28,817 between 2002 and 2008, a period of major improvements in the agency's overall capacity.

a. Many other smaller NMSs have fewer than five employees, but those agencies did not participate in the survey.

(Zillman 1999)—especially in developing countries, where available funds have fallen far short of those needed to maintain the standards of infrastructure and service provision.

The funding gap (for both developing and developed countries) has historically been partially filled by user groups, such as the civil aviation industry—either through charges based on volume of service, incremental fees for user-specific services, industry-specific taxation measures, or direct percentage-based contributions to the cost of the total operation (Zillman 1999).

Since the 1980s, with the widespread introduction of user fee regimes for many services formerly provided by the government, several governments have experimented with alternative approaches to funding the operations of their NMHSs. But the monopoly on the basic observation infrastructure makes it difficult for an NMS to operate a fully privatized entity subject to competition law. The overall level of funding varies greatly, reflecting the wide range of national circumstances (size, geography, population density, history, stage of development, economic circumstances, and government policies on service provision) (see box 5.2).

Box 5.2 Funding Levels and Budgetary Pressures

NMS Funding

The total estimated annual budget for National Meteorological Services (NMSs) globally, including satellite operations, is significantly more than US$10 billion. For instance, the National Research Council assessment (NRC 2012) argues that total funding of the weather enterprise in the United States in 2012 ranged from US$8–US$10 billion, split equally between the federal and nonfederal levels. Most NMSs are funded in the range of 0.010–0.050 percent of gross domestic product, with a global average of 0.012 percent. Per capita spending on NMSs ranges from virtually zero—less than US$0.10 for at least one NMS in every World Meteorological Organization (WMO) region—to almost US$13.00. The average for developed countries is about US$3.50 per capita of national population, and the average for least developed countries is about US$0.25 (Zillman 2003).

Fiscal Pressures

Of the 30 NMSs responding to the Environment Canada 1997 survey, 50 percent considered the stability of their services to be threatened by budget pressures forcing changes in service provision. Overall, the majority of NMSs surveyed (over 70 percent) considered that budget restraint and its fallout were serious obstacles to fulfilling their mission in a sustainable fashion, resulting in a greater emphasis on commercialization (Jean et al. 1999). In a 2000–01 WMO survey, more than 50 percent of the NMSs responding reported cost-recovery levels of less than 10 percent of their cost of operation, but a small number (under 5 percent) had cost-recovery levels in excess of 50 percent. Cost recovery is most commonly practiced in Europe, and the cost-recovery levels are higher there (around 20 percent). In the 2011, WMO survey, 89 percent of the NMSs responding indicated that the overall level of government funding was either a very significant or a significant issue (WMO 2012).

Funding Sources

According to the 2000–01 WMO survey, the vast majority of NMSs (over 60 percent) operated as organizations for public good and received most of their funding (about 80 percent) in the form of direct government appropriations. About 81 percent of NMSs reported that they received government funding for the priority areas of warning services and public weather

box continues next page

Box 5.2 Funding Levels and Budgetary Pressures *(continued)*

services, according to the 2011 WMO survey. However, 14 percent and 21 percent of the NMSs indicated that warning services and public weather services, respectively, were provided commercially or on a cost-recovery basis (WMO 2012).

A Need for Appropriate Operating Models

Notwithstanding the roles of the private sector, academia, and others, NMHSs are responsible for the basic national meteorological and hydrological observation networks. They deliver forecasts and analyses and issue meteorological and hydrological warnings to the government and the public.

The dominant role of NMHSs as national weather, climate, and hydrological service providers creates opportunities to improve economic performance in various weather-, climate- and water-sensitive sectors. Conversely, there is also the threat that they will exploit their near monopoly on weather, climate, and hydrological data to limit opportunities for others to develop products and services. Thus, NMHSs' control over such data is a major source of tension among the various actors seeking to provide commercial weather services in both developing and developed countries.

Efforts to modernize NMHSs have tended to focus on up-front investment in equipment, but not on the resolution of these institutional issues. As a result, the improvements often prove unsustainable. Appropriate operating models are needed—that is, models that cover (a) what NMHSs do, (b) how NMHSs are funded and managed to sustain their core functions and responsibilities for basic forecasts and public safety, and (c) how their relationships and partnerships with other actors are defined to enhance cooperation rather than suppress innovation.

Given the financial realities within the public sector, it is natural to ask the following questions:

- Are there opportunities in developing countries for greater cooperation between the public and private sectors and academia?
- Will partnerships of this kind lead to greater sharing of data and information?
- Can a win-win situation be created that fulfills the public sector responsibility to help the economically disadvantaged (such as smallholder farmers in Africa), while meeting the needs of large enterprises (such as insurance, agribusiness, and hydropower companies)?
- Is there a sustainable financial model for NMHSs?
- Can business operating models of NMHSs in developed countries be adapted for use in the least developed countries?

In trying to answer these questions, one should keep in mind the weather and climate services value chain, which highlights the different roles of NMHSs (figure 5.1). The most critical responsibilities are to provide basic forecasts and warnings to protect society from the adverse effects of severe weather—a

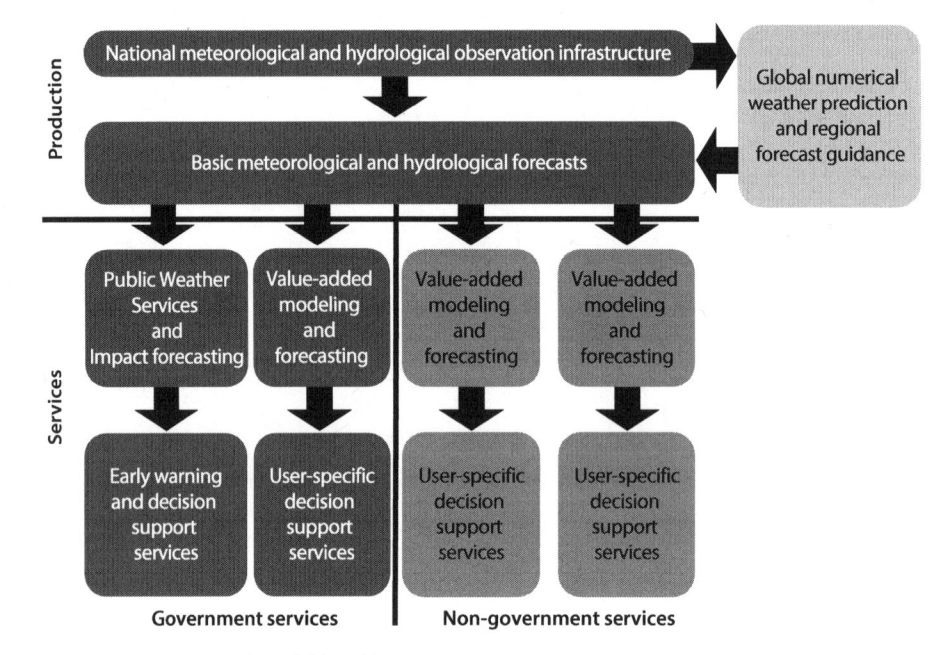

Figure 5.1 Weather and Climate Services Value Chain

```
Production:
  National meteorological and hydrological observation infrastructure
        ↓                                                    → Global numerical
  Basic meteorological and hydrological forecasts               weather prediction
                                                            ←   and regional
                                                                forecast guidance

Services:
  ┌─────────────┬─────────────┬─────────────┬─────────────┐
  │ Public      │ Value-added │ Value-added │ Value-added │
  │ Weather     │ modeling    │ modeling    │ modeling    │
  │ Services    │ and         │ and         │ and         │
  │ and Impact  │ forecasting │ forecasting │ forecasting │
  │ forecasting │             │             │             │
  ├─────────────┼─────────────┼─────────────┼─────────────┤
  │ Early       │ User-       │ User-       │ User-       │
  │ warning     │ specific    │ specific    │ specific    │
  │ and         │ decision    │ decision    │ decision    │
  │ decision    │ support     │ support     │ support     │
  │ support     │ services    │ services    │ services    │
  │ services    │             │             │             │
  └─────────────┴─────────────┴─────────────┴─────────────┘
     Government services      |   Non-government services
```

Note: Users are governments, households, and businesses.

responsibility that should be supported by governments. NMHSs can also use global models and data and the national observation infrastructure to provide specialized services to government agencies and individual businesses. The latter may be either a public sector responsibility or a commercial opportunity to derive revenue from nongovernment sources.

Demand Side: The Users

Starting with the end of the weather and climate services value chain, different types of users have different types of needs. What they ultimately receive can be separated into broad two groups: (a) basic services (that is, public communication of general weather forecasts and warnings) and (b) value-added services (which are tailored to specific users in the public and private sectors). Overall, the value of these services depends on how well the recipients can use the information to meet their needs (see box 5.3).

Basic and Specialized Services for the General Public and Government

Basic services include routine daily forecasts and longer-range outlooks, as well as severe weather and hydrological warnings based on nowcasts and very short-range forecasts. In response, the government is expected to make decisions and act (for example, coordinating evacuation in case of expected flooding); to take steps to minimize loss of life; and to adopt measures to reduce the risk of damage to public infrastructure. When warnings are issued, the government entities responsible

Box 5.3 Benefits of National Meteorological and Hydrological Services

National meteorological and hydrological services benefit most sectors of society:

- *Households and individuals* plan their activities according to day-to-day weather forecasts.
- *Governments* take appropriate measures to minimize loss of life (for example, through France's early-warning system for health) and to reduce the risk of damages to public infrastructure during severe weather conditions.
- The *transportation sector* (air, land, and water) ensures efficient and reliable transport of people and goods by consulting meteorological forecasts.
- The *water sector* manages water resources according to the availability of rainfall and risk of flooding.
- The *agricultural sector* reduces its vulnerability to severe weather conditions in the short and long run by adapting agricultural practices to meteorological and climatological conditions.
- The *energy sector* relies on weather and water forecasts and climate predictions for optimizing production, plant management, and investment decisions. Weather and climate information is particularly important for power generation based on fluctuating renewable energy (such as wind power or solar radiation). Hydrological forecasts are important for managing dams.

for national, provincial, and local responses to meteorological and hydrological hazards should closely cooperate with one another to ensure timely action.

Specialized and Value-Added Services for the Commercial Sector

Many economic sectors increasingly depend on meteorological information for safe and efficient operation. They often require information specifically tailored toward their sector or, in some cases, certain businesses. Such information may include altitude conditions for airlines or more geographically specific and longer-term information for farmers. In addition, because many businesses typically maintain low inventories of goods to minimize costs, supply chains are more susceptible to transport network disruptions, which can seriously affect the availability of essential commodities, from food to heating oil. Conversely, if provided with adequate lead time, such agencies can increase stock and resources in anticipation of requirements. Tailoring meteorological and hydrological information to the specific needs of a particular business makes that information exclusive and provides the basis for value-added payment. But whether payment is actualized may depend on the nature of the business. Products tailored to large-scale agricultural enterprises would normally be provided for a fee, whereas similar information provided to smallholder farmers might be provided at no cost as part of the specialized services for the public.

Do the benefits outweigh the costs? Overall, the evidence suggests that well-functioning NMHSs—when matched with appropriate resources—provide substantial socioeconomic benefits, well in excess of their costs (see box 5.4 technical insight).

> **Box 5.4 Technical Insight: Quantifying Socioeconomic Benefits**
>
> Although determining the costs associated with weather and climate services is relatively straightforward, quantifying the socioeconomic benefits is more difficult. A nonmarket valuation is needed, given that the revenue gained from providing weather and climate services is only one factor governing their value. Despite these challenges, the studies available so far indicate that the benefits exceed the total costs.
>
> For the United States, Lazo, Morss, and Demuth (2009) determined a cost-benefit ratio of 1:6 on the basis of surveys of 1,465 households addressing the routine use of weather information and excluding the added benefit of warning systems, which would increase the benefits. In Switzerland, Frei (2009) determined a cost-benefit ratio of 1:5 for households and the agricultural and energy sectors. In a more recent study investigating the benefits for the road transport, rail transport, aviation, and energy sectors, Bade, von Grünigen, and Ott (2011) estimated the value of weather services in Switzerland to range between US$100 million and US$121 million per year in contrast to total annual costs of US$85 million. Given that benefits to individuals, as well as to the agricultural, tourism, insurance, and building industry sectors, were neglected in this study, the actual benefit is assumed to exceed the published value.
>
> For the European program GMES (Global Monitoring for Environment and Security), in which National Meteorological and Hydrological Services (NMHSs) play an important role, cost-benefit assessments show that earth observation services will reveal an annual benefit of €10 billion by 2030.
>
> For all developing countries, Hallegatte (2012) estimates that the total benefits of upgrading all hydrometeorological information production and early-warning capacities to the standards of developed countries would be between US$4 billion and US$36 billion per year globally, with benefit-cost ratios of between 4:1 and 36:1.

Supply Side: The Providers

What is involved in providing weather, climate, and water services? The answer can be broken down into several "links" or components of the supply chain. Most, if not all, weather, climate, and water services—whether provided by NMHSs or the private sector—depend on the national observation infrastructure. These data are used in stand-alone data sets for time-series analysis or are more usefully assimilated into nowcasting and forecasting systems (see chapter 3). National observations are combined with other data—such as satellite and surface observations from other countries' networks—to create the best possible numerical weather prediction products, which are in turn shared with all NMHSs. More extensive national observations may also be assimilated into local area models, which obtain their initial conditions from the global numerical weather prediction systems and then provide more rapidly updated forecasts at regional and local scales.

The value of weather and climate services hinges on the composition and accuracy of the underlying national meteorological and hydrological observation networks, which are built on various components. Adding another component potentially increases the overall accuracy, but the marginal benefit is likely to decrease for already well-established observation systems. In addition, implementing an additional component may be associated with considerable operating and maintenance costs, which need to be traded off against the expected benefit. Such cost-benefit analyses in countries with advanced observation systems may help optimize the observation infrastructure in developing countries, where financial resources are particularly constrained, but also where observation infrastructure needs extensive investment for modernization.

Thereafter, the provision depends on the nature of services. Basic services require global models and general forecasts and then public communication of the forecasts and meteorological and hydrological warnings. The nature of value-added services varies. They still draw primarily on national networks, but they may, in some cases, include (a) operating dedicated networks to provide additional observations (for example, more detailed local information for agricultural management); (b) developing and implementing more sophisticated modeling tools; and (c) incorporating meteorological information with additional related information to provide value-added services to help with decision making (for example, about how the particular sector responds to the weather information for disaster response planning). In some cases, it may simply be a matter of packaging the basic forecast in a way that immediately provides specific users with what is needed.

How Economic Characteristics Fit In

In weighing how to divide up responsibilities among national and local governments, academia, and private companies and whether and how users can and should be charged for the services, one may find it helpful to draw on economic theory (Freebairn and Zillman 2002). Keep in mind that public goods have two class characteristics: nonrival and nonexcludable.

National Observation Infrastructure and Data Assimilation: Nonrival but Excludable

Given the natural monopoly characteristics of a single national network—high capital costs and fixed costs—a private market would not be competitive because it would make no sense to have more than one provider. Plus, the economically efficient price (the marginal cost of production) of these data is zero. But potential data users can be excluded, which might be an issue, because once paid-for data are released, it can be hard to limit their circulation.

Basic Forecasts and Warnings: Nonrival and Nonexcludable

The nonrival and nonexcludable characteristics make it inefficient to ask users to pay for basic forecast services because the cost of excluding users to avoid free riders would be prohibitive and the marginal cost of serving another user is zero.

Value-Added Services: Rival and Excludable
Different users need specific information, modeling, or bundling. Therefore, services could be funded by user fees. But providers need to consider certain factors:

- Some of the benefits are external to private companies that would buy the service (for example, a safe construction site would benefit people near it) so governments may at least need to regulate to ensure that those companies get all the necessary information.
- Information failures may occur, whereby potential users of data either do not know or miscalculate how useful the data would be. There is considerable economic evidence of people underestimating the value of information about risk. Such underestimations are particularly likely to occur where the market for data is underdeveloped (for example, among smallholder farmers). In such cases, it may make sense for the government (typically, the responsible ministry) to be involved.
- In some cases, the value-added output may itself contribute to more effective or efficient provision of a public or otherwise publicly provided good (for example, treating roads for icing might help in a disaster response).
- There may often be a monopoly buyer for a particular set of information (for example, another government department). In that case, the services cannot sensibly be competed and should be part of the public provision of basic and value-added services.
- Barriers to entry may or may not exist for other service providers. For example, if data are freely available, others may provide value-added services competitively.

These economic arguments suggest that governments support NMHSs for good reason. There are many elements of public goods, which is why NMHSs have always relied on government funding. However, clearly roles exist for other entities to provide weather, climate, and water services. Thus, we need to carefully weigh the appropriate model and role for NMHSs, along with how they should or can be funded. And to do that, we need to examine (a) the scope of NMHSs services, (b) access to and pricing of NMHSs data (and services), and (c) possible management models of NMHSs.

Scope of NMHSs Services
How broad is the scope of NMHSs services? The traditional role of NMHSs emphasizes producing and disseminating basic meteorological and hydrological forecasts and warnings (see chapters 3 and 4). NMHSs may also play a role in communications. Depending on national policy, they may be responsible for providing forecast information and warnings directly to the public through their own media outlets and through private or public sector media. For example, the U.S. NWS provides such services through its own website, but it relies on commercial and public television and radio stations to broadcast weather information. Government regulation ensures that warnings are

carried by all broadcasters in a common format. In the United Kingdom, the Met Office is contracted by the British Broadcasting Corporation to provide forecasts on air; in contrast, the independent networks purchase information but use their own presenters on air. In China, the China Meteorological Administration operates a broadcast meteorological service in partnership with a state-owned company that employs a staff of more than 800 and supports at least 80 separate television channels nationally and internationally. This partnership combines both a public service role and a commercial service.

The role of NMHSs in providing value-added services is twofold:

- NMHSs can be single providers of value-added services in governments if these services cannot sensibly be competed. This role usually depends on the particular operating model. If NMHSs are fully funded by their parent ministries, their services are usually provided at no additional charge. In other cases, government agencies may charge each other for services. In Nepal, the government requires ministries to compensate each other for data. In the United Kingdom, government agencies may share costs, compete, or single-source weather services—although typically they need to cooperate—particularly if governments need help considering and defining how meteorological and hydrological services could be used.
- NMHSs can be general providers of other value-added services. In this case, the NMHSs may compete with other providers.

Access to and Pricing of NMHSs Data and Services

Following economic and practical considerations, a liberal approach toward data policy with the free-of-charge provision of infrastructure data and products is very likely to generate socioeconomic benefits (Weiss 2002). The rationale is that basic meteorological data and services are characterized as public goods; hence, consumption is nonrival. It follows that the marginal cost of supply for additional consumption is close or equal to zero if the information is readily available. Furthermore, costs of exclusion from consumption (that is, controlling proliferation of charged data) are too high, if not impracticable. As a result, state-funded and free basic data access is likely to have a positive effect on social welfare.

In contrast, value-added and tailored special services for particular customer groups have features of a private good (rivalry in consumption and low costs of exclusion). In this case, it is relatively straightforward to identify the additional inputs required for producing refined products and to charge users a fee equal to marginal costs of additional inputs (Freebairn and Zillman 2002; Zillman and Freebairn 2001).

Attitudes toward data liberalization vary considerably. The Netherlands, Norway, the United States, and to some extent Iceland and the United Kingdom have followed a liberal data policy. However, the majority of European NMHSs charge for meteorological data and services.

Options for NMHSs Management Models

What are the options for managing the elements of NMHSs that remain within their scope? They could be a government department, a public enterprise, or even a public-private partnership (PPP). Each has implications for funding options, management incentives, and management autonomy:

- *Funding options.* Such options exist, but it should be noted that no NMHSs have managed to raise more than 20–40 percent of their revenues from nongovernment sources. Therefore they are never going to be viable as fully private enterprises without a subsidy.
- *Management incentives.* These incentives allow autonomy, remuneration, and efficiency.
- *Management autonomy.* Such autonomy includes more flexibility with respect to personnel policies.

Governments are also seeking ways to improve efficiency and reduce the costs of public services. In this context, operating models play a particularly important role, although they must face the challenge to find a balance between management efficiency and political control of service provision (Broussine 2009). Methods have varied from the outsourcing of public services to the transfer of public functions to more independent government entities that operate at arm's length from their originating departments. Services associated with public safety have remained largely within the control of central, regional, or local government departments.

Despite having responsibility for protecting lives and livelihoods, some NMSs have become targets for quasi-commercialization (Freebairn and Zillman 2002). However, transformation processes have not always produced the desired results, and as more and more developing countries change the way the civil services operate, there is a risk that some of these problems will be perpetuated.

Fortunately, there is now a body of knowledge about business practices in the public sector. This knowledge is based on the experiences of a growing number of countries that have reformed their civil services and, as a consequence, their NMHSs.[5] It shows that the degree of financial, managerial, and personnel autonomy varies considerably among the entities providing public goods and services (Van Thiel 2009). Of course, that autonomy makes it difficult to define a general framework for organizational types (Gill 2002; Pollitt et al. 2005), plus there is no international consensus regarding the terminology.

Even so, Van Thiel (2001) and Greve, Flinders, and Van Thiel (1999) have identified an organizational structure for public sector bodies common to several governments. It features six types of entities, each characterized by an increasing degree of autonomy:

- Government departmental unit (hierarchical unit that operates under direct control of the minister)
- Government agency (quasi-autonomous part of the department)

- Public body (operating at arm's length from the central government)
- Voluntary or charity organization (bottom-up body providing public services)
- Privatized company (former state-owned company that is fully privatized)
- Contractor (private entity contracted by the government for public service provision)

Which model a country chooses depends on the political, economic, and societal landscape, including the national legislative framework. Other criteria may serve as a rough guideline within the decision-finding process for an operating model[6]:

- *Degree of government involvement.* To what degree is the government involved in providing public services? Is the government involved only in strategic orientation (such as appointing the governing board or the executive board), or is it allowed to intervene in operational activities?
- *Equity.* To what degree is the entity owned by the government? Is it state owned, or does the government hold minority interests? This criterion also addresses the question of the degree to which the government budget finances the organizational entity.
- *Internal autonomy.* Is the organization enabled to issue rules? Does it have a separate accounting system? How much autonomy does it have with respect to personnel policy and management?

Operating Models for NMHSs Services

Using the six categories as a starting point, we will now adjust them to correlate with the operating models evident in NMHSs today (figure 5.2). We will add one more type—the state-owned enterprise (SOE), which replaces the contractor category and can be slotted between public bodies and privatized companies—and drop the voluntary or charity category. In practice, most NMHSs are departmental units, agencies, or public bodies.

Model I: Departmental Unit

Departmental units are hierarchical units operating under direct control of the corresponding minister, who has direct political control and ministerial responsibility for the unit's performance. Examples of departments that have such units are the departments of justice, economics, or defense (Greve, Flinders, and Van Thiel 1999). The units are financed by the state budget, are typically single-purpose bodies, and deliver noncommercial services to citizens or support other public sector bodies. They operate under public law and do not have their own legal personality (Gill 2002). A good example is the U.S. NWS (see box 5.5).

Model I is the most straightforward model to implement if public sector financing is sufficient. The risk of this approach is that, because there are no alternative sources of revenue, services will be reduced if public spending is reduced. In many countries, cuts in public sector financing mean that the NMHSs

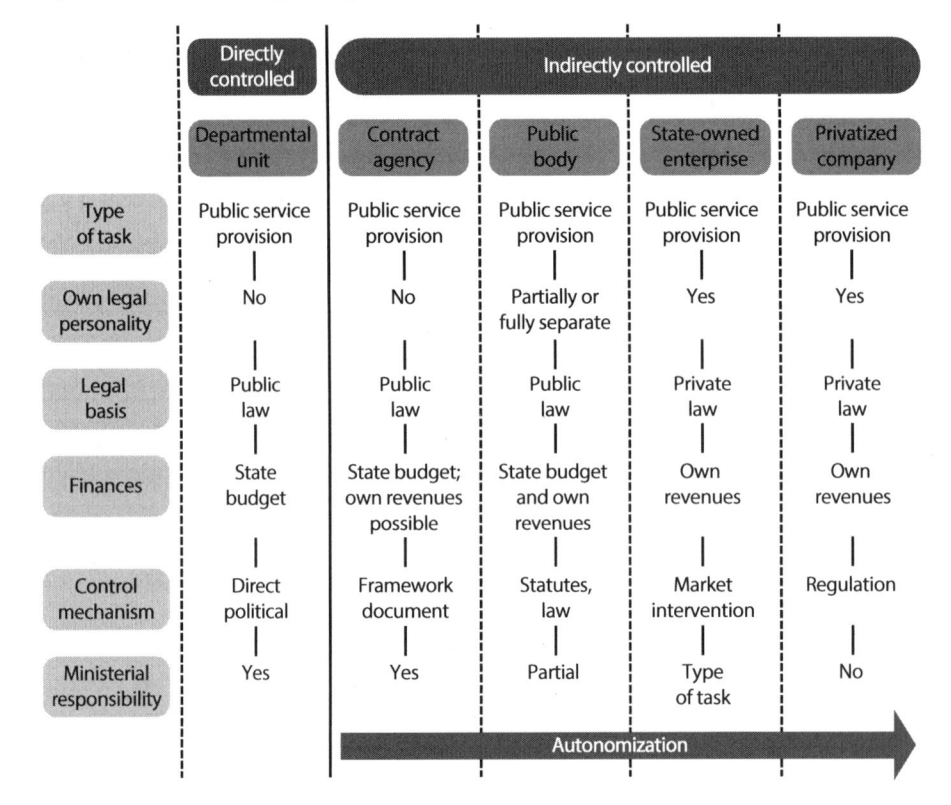

Figure 5.2 Five NMHSs' Operating Models

	Directly controlled	Indirectly controlled			
	Departmental unit	Contract agency	Public body	State-owned enterprise	Privatized company
Type of task	Public service provision	Public service provision	Public service provision	Public service provision	Public service provision
Own legal personality	No	No	Partially or fully separate	Yes	Yes
Legal basis	Public law	Public law	Public law	Private law	Private law
Finances	State budget	State budget; own revenues possible	State budget and own revenues	Own revenues	Own revenues
Control mechanism	Direct political	Framework document	Statutes, law	Market intervention	Regulation
Ministerial responsibility	Yes	Yes	Partial	Type of task	No

Autonomization →

Sources: Adapted from Gill 2002; Greve, Flinders, and Van Thiel 1999.
Note: The first four models can work with privatized companies as public-private partnerships.

Box 5.5 Model I Example: U.S. National Weather Service

The U.S. National Weather Service (NWS) is a departmental unit that is core funded to provide all its products and services at public expense for the public good. This approach has helped to develop a strong private sector that does not directly compete with the NWS, thus maintaining a balance between the public sector's role in providing basic information for public safety and economic security and the need to sustain specialized, value-added services in the private sector (NRC 2003). That balance permits a clear distinction among public goods, private goods, and mixed goods.

In the United States, the weather enterprise has three parts: (a) the public sector in the form of the NWS, which is responsible for protecting life and property and for enhancing the national economy; (b) academia, which is responsible for advancing the science and for education; and (c) the private sector, which is made up of weather companies, meteorologists working for private companies or as private consultants, and broadcast meteorologists and is responsible for creating tailored products and services for individual clients and for working with the NWS to communicate forecasts and warnings that may affect public safety (NRC 2003).

are competing with other government priorities (such as primary and secondary education, public health, and national security). Poor public financing creates a downward spiral for NMHSs that results in reduced staffing, an inability to maintain observation networks, a limited capacity to innovate, low organizational incentives, and poor service delivery—which in turn lead to further neglect and an inability to meet the need for adequate warnings of extreme events.

Moreover, for various reasons, an inherent tension exists between the public, private, and academic sectors, as the NRC (2003) points out:

- Each sector contributes in varying degrees to the same activities—data collection, modeling and analysis, product development, and information dissemination—thus making it difficult to clearly differentiate their roles.
- Each sector has its own philosophy of sharing data and models with the other sectors and with the general public.
- Advances in scientific understanding and technology permit new user communities to emerge and change what the sectors are capable of doing and want to do.
- Not all members share the same expectations and understanding of the proper roles and responsibilities of the three sectors.

In the United States, reducing friction among these sectors is important for the smooth functioning of the weather enterprise. About 40 countries continue to use this model (John Zillman, personal communication, e-mail, January 5, 2013)—and of those that do, many are in transition or have recently transitioned to one of the other models (for example, Kenya, Tanzania, and Uganda).

Model II: Contract Agency

Contract agencies are quasi-autonomous parts of government departments or ministries. These entities have more autonomy than departmental units, but they still have a strong hierarchical and financial relationship with the parent department and the minister. Consequently, the minister is politically responsible for the entity's performance. Examples of agencies are the Danish Patent Office, the Netherlands Service for Immigration and Naturalization, and the Swiss Federal Office of Meteorology and Climatology (see box 5.6). The activities of contract agencies are determined by a framework document, which defines performance and budget guidelines (Greve, Flinders, and Van Thiel 1999). In general, contract agencies are financed by the state budget. However, depending on the organizational structure, they may generate their own revenues (for example, through user fees). Similar to the departmental units, contract agencies operate under public law and do not have their own legal personality (Gill 2002).

Model III: Public Body

Public bodies refer to the entities that operate at arm's length from the central government. As a consequence, they face less political and hierarchical influence and have more operational and managerial freedom. The control mechanism is

Box 5.6 Model II Example: MeteoSwiss

In the late 1990s, the Swiss Federal Council implemented the so-called four circles model for its public services. The goal was to encourage the outsourcing of public services from the central administration to entities with a higher degree of economic and operational freedom in hopes of achieving greater effectiveness and efficiency (SFC 2006).

The result is a system where political influence decreases while market dominance increases the farther out the entity is located in the circles model (figure B5.6.1). The first, innermost circle refers to the entities of the departmental units that are closely related to politics. The second circle includes the entities that operate under a global budget and a performance mandate defined by the government. The third circle is composed of public bodies. And the fourth circle refers to the state-owned enterprises (SOEs; SFC 2006; Steiner and Huber 2012).

Figure B5.6.1 The Swiss Federal Council's Four Circles Model

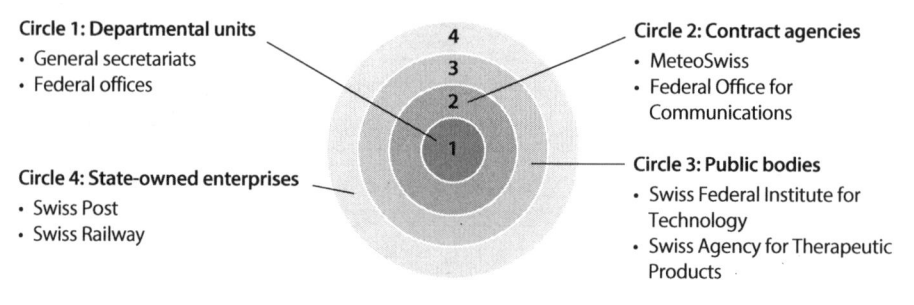

Circle 1: Departmental units
- General secretariats
- Federal offices

Circle 2: Contract agencies
- MeteoSwiss
- Federal Office for Communications

Circle 4: State-owned enterprises
- Swiss Post
- Swiss Railway

Circle 3: Public bodies
- Swiss Federal Institute for Technology
- Swiss Agency for Therapeutic Products

Sources: Adapted from SFC 2006; Steiner and Huber 2012.

The four circles model is not static and allows organizational transformation and autonomization processes within the framework. In Switzerland's case, these processes are closely related to the national political discourse, as the following example of the proposed transformation process of the Federal Office of Meteorology and Climatology (called MeteoSwiss) shows.

Traditionally, MeteoSwiss has been a contract agency (circle 2). But in 2008, a plan was proposed to reorganize it into a public body (circle 3). The impetus for this change was the general trend toward free access to government data, which would lead to a loss of revenue for MeteoSwiss. To cope with these changes, the government formulated a new strategy that was based on the following key elements:

- Continuation of monitoring of weather and climate
- Forecasting of weather and climate on all time-scales
- Assurance of early warnings of the population to mitigate weather-related natural hazards
- Free access to all relevant basic data
- Formulation of a legal entity with its own account and increased economic freedom

A new law on meteorology, which was a prerequisite for the reorganization, was elaborated on the basis of these key elements. However, a public submission of the law—which is

box continues next page

Box 5.6 Model II Example: MeteoSwiss *(continued)*

mandatory in Switzerland—and political reactions revealed three areas of conflict: (a) the degree of private activities (including public relations and sponsoring), (b) the working conditions of civil servants, and (c) a reluctance to decentralize the services. As a result, in October 2012, the Swiss Parliament decided to keep MeteoSwiss as a contract agency (model II), thereby retaining the government's supervisory and fiscal responsibility for the organization.

---

based on statutes or laws. And owing to the greater autonomy, both the central government and the public body are responsible for performance. Public bodies are usually responsible for regulatory matters and for managing public universities. Some entities are financed by the state, whereas others rely on their own revenues (such as levies) on top of the state budget. Public bodies are based on public law and, depending on their level of autonomy, have a partially separate or fully separate legal personality (Gill 2002; Greve, Flinders, and Van Thiel 1999).

Governments usually refer to the additional discretionary funding mechanisms of public bodies as commercial, although the exclusivity normally associated with commercial activities is lacking. Even if the government is still responsible for providing public services, the underlying goal is to shift the complexity of service provision to the public body. This organization type is expected to enhance professional management and to place services closer to citizens (Pollitt et al. 2005).

Over the past 25 years, European NMHSs have experimented with various forms of commercialization in public bodies. As a result, they are almost always organized as entities that work at arm's length from the government (Van Thiel 2012). The enthusiasm for developing public sector commercial services has been strongest in France, the Netherlands, the Scandinavian countries, and the United Kingdom (see box 5.7), with varying degrees of success. And many African countries are in the process of transforming meteorological departments into public bodies in the hopes of creating more flexible organizations with the capacity to make strategic and tactical decisions on their own, especially on budgeted resources.

This model or some variation occurs in most NMHSs, but it is difficult to implement unless it is part of a governmentwide reform. The reason is that government entities that are not used to paying for services from other government departments will likely refuse to pay for meteorological and hydrological information, even though it may be essential for their operations. NMHSs must adopt good accounting methods to properly monitor and demonstrate the cost-effectiveness of these services. And guidelines on accounting, cost centers, and management practices are needed, which could be developed with the World Meteorological Organization (WMO).

One consequence of this approach can be an overall reduction in core spending on infrastructure with the aim of making up shortfalls from contract funding. The problem is that contract funding is usually short term, and ongoing basic infrastructure costs may not be met. There is also a need to dedicate many staff

Box 5.7 Model III Example: U.K. Met Office

The U.K. Met Office's operating model distinguishes between two types of customers: (a) central government bodies requiring services that cannot sensibly be competed and (b) services provided on a commercial (usually competed) basis to customers both inside and outside government. This distinction helps the Met Office demonstrate its compliance with competition law in relation to possible cross-subsidy (Met Office 2007).

The main customers for noncompeted services to the central government are the Ministry of Defence; the Department of Environment, Food, and Rural Affairs; and the Public Weather Service Customer Group. In these cases, prices are to be set at a level consistent with Treasury guidance on the cost of capital for interdepartmental services. The Met Office maintains customer-supplier agreements with all relevant customers. These agreements clearly define the outputs and associated costs and, where possible, provide incentives for cost reduction or service enhancement. In contrast, profit margins for business, which is either competed or capable of being competed, are dictated by market conditions. This characteristic also applies to competed services to government departments (Met Office 2007).

Developing the appropriate policy and legal framework is essential. In the United Kingdom, the Met Office's activities are defined within the Government Trading Funds Act 1973, as amended, and the Meteorological Office Trading Fund Order 1996, with subsequent amendments. A trading fund is expected to make a return on capital employed, which is a dividend for the Treasury.

If this practice is used elsewhere, it is recommended that any return on capital employed be used within the NMHSs to invest in new services or to provide staff incentives rather than be returned to the Ministry of Finance. Unfortunately, a common practice in many developing countries is to return revenue generated directly to the Treasury and not to contribute to sustaining or improving services. This lack of reinvestment often results in a deteriorating, overpriced service, which engenders growing customer disaffection and a decline in political support.

---

members to soliciting and managing contract work, with a greater emphasis on development, sales, and marketing rather than on forecasting. In small NMHSs, dedicating so many staff members to these functions may be difficult to achieve. In addition, a problem may arise during the transition or even before of how to handle data from the observation system database, which most NMHSs consider to be their only asset. As a result, they tend to restrict access to these data. The net effect is that fundamental questions about climate adaptation may go unanswered for fear that this asset will be exploited by others at the expense of the NMHSs.

Model IV: State Owned Enterprise

SOEs—a critical element of public service provision in a significant number of countries (Gill 2002)—are characterized by an even higher degree of operational and managerial autonomy than model III entities. They operate under private law (that is, law that regulates corporations and private companies); they have their own legal personality; and their stocks are held and controlled by the government. However, there is no uniform definition of *governmental control*. For instance, the United Nations Conference on Trade and Development defines *control* as a situation in which the government has 10 percent or more of the voting power or is the largest single shareholder (UNCTAD 2011).

SOEs are often found in the fields of postal services, railways, or power supply. Similar to public bodies, SOEs have to fulfill the tasks assigned by the government and are controlled through market interventions or by the buying or selling of shares. In addition, ministerial responsibility is usually limited to strategic and financial decisions. SOEs are usually financed by their own revenues (Gill 2002; Greve, Flinders, and Van Thiel 1999; Van Thiel 2001), even though some enterprises are subsidized by the government—especially investment-intensive public services, such as railway infrastructure maintenance.

Model IV applies to NMHSs that are encouraged or required to compete directly in competitive commercial markets. One of the first model IV NMHSs occurred in New Zealand (see box 5.8). In such a situation, the National Meteorological Service acts as a wholesaler of basic meteorological and hydrological data and products that can be purchased by the retail commercial or discretionary funded service. All but the core public service would pay and transfer the costs to their customers. This model also applies to those countries that have decided to contract out public weather services to the private sector or have created a government-owned commercial enterprise. In this case, the government has defined its requirements for a weather service, and the enterprise fulfills its contractual obligations to discharge this public responsibility in the same way that a public body would. The leadership of this enterprise would participate on behalf of the country in international bodies such as the WMO.

Box 5.8 Model IV Example: New Zealand MetService

New Zealand was one of the first to adopt the state-owned enterprise (SOE) approach. The New Zealand Meteorological Service had operated as an agency of the Ministry of Transport. But in a series of reforms in the science and transport sector, it was divided into MetService and the National Institute of Water and Atmosphere, which started to operate in July 1992.

MetService was established as an SOE to operate as a commercial business, providing New Zealand's national weather service. It has a wholly owned subsidiary, Metra, which provides international commercial services. The National Institute of Water and Atmosphere, which is responsible for climatic issues and scientific research, was incorporated into the Crown Research Institutes. It is also commercially active, but unlike MetService, it is not required to ensure a dividend (Steiner et al. 1997).

Although there was considerable mistrust of MetService when it was first created, it has proved that an SOE, operating as a commercial company, can provide public services and discharge the nation's international obligations, if vested by the government to do so, and can provide a return on investment to its shareholders.

One caveat here is that asking NMHSs to compete in competitive commercial markets raises a number of issues regarding the monopoly that NMHSs hold on basic hydrometeorological and related data and fair competition. NMHSs are data monopolies that enable the effective reuse of public information by individuals and the private sector. However, without a clear separation of their public sector tasks from their commercial interests, it is easy to see how NMHSs may inadvertently—or otherwise—misuse their dominant position in the commercial market for hydrometeorological and related information.

Model V: Privatized Company

If an SOE is further autonomized, the result is a privatized company. This transformation process is common in highly competitive sectors such as telecommunications or air transport (Greve, Flinders, and Van Thiel 1999; OECD 1995). Similar to the SOEs, privatized companies operate under private law and have their own legal personality. But unlike the previous models, they are fully responsible for their performance. They operate freely in the market and generate their own revenues. Even though direct political control decreases through the privatization process, economic activities are controlled to a certain extent by regulations (Hughes 2003).

Privatization typically refers to the transfer of ownership and control of government or state assets to the private sector. Privatization may also include the transfer of firms or operations. It is widely accepted that governments should not own companies in the competitive sector (such as financial services or industry). However, the situation is different for public services (such as energy or transport), where opinion regarding privatization diverges greatly (OECD 1995). An example of privatization in the field of NMHSs occurred in the Netherlands where the Royal Netherlands Meteorological Institute (KNMI) privatized its commercial interests into the Holland Weather Services (HWS; see box 5.9), with KNMI reverting to fulfill the role of a government entity closer to the departmental model (model I). So, in fact, full privatization does not appear to be a real option for NMHSs, whose primary responsibility is to the public and government, but is reserved for spin-offs of public sector bodies that operate entirely within the commercial realm.

Public-Private Partnerships

Even further autonomization would yield yet another approach for NMHSs: partnering with the commercial sector or nonprofit sector to create an entity that can broker information and data exchange between the NMHSs and the private sector. Such a partnership may also be a conduit for acquiring observation stations to strengthen national networks while enhancing observations of highest value to the commercial sector. This concept, known as a *public-private partnership* (PPP), provides an alternative to conventional procurement and contracting. The aim of a PPP should be to provide more efficient and qualitatively better services as a result of bundling and generating synergies of public and private

Box 5.9 Model V Privatized Company: Holland Weather Services

In 1999, all commercial activities of the Royal Netherlands Meteorological Institute (KNMI) were privatized into the Holland Weather Services (HWS). This decision was preceded by increased tensions between KNMI and its private competitors because of KNMI's intensified activities in the commercial sector.[a] The shares of HWS are held by Weather News Inc. (75 percent), a Japanese meteorological multinational, and by a Dutch entrepreneur (25 percent). HWS now competes on the national market for commercial services, including media contracts, Internet services, contracts to the shipping industry, and forecast packages for specific customers such as farmers (Pollitt et al. 2005).

KNMI, which serves as a national knowledge center, covers noncommercial services. One example is the meteorological service for the aviation industry, which is considered a public (or noncommercial) service because it is closely related to national security (Pollitt et al. 2005). KNMI also maintains monopolistic control over the national observation infrastructure needed for weather forecasts and warnings. But given that KNMI no longer has any contracts with the media, private companies must ensure that weather warnings by KNMI are delivered to the media in time (Pollitt et al. 2005).

a. In Norway, the case of Storm Weather Center AS and the Norwegian Meteorological Institute illustrates potential tensions between state and private enterprises regarding commercial activities.

---

resources. And there is considerable scope for developing PPPs to deliver weather and climate services.

PPPs involve long-term cooperation between at least one public and one private partner for the joint management of complex projects in which there is an exchange of know-how and a sharing of risks in fulfilling public tasks. PPPs can be viewed as a third path between administrative reforms and total privatization (Lienhard 2006). Cooperation is based on a contractual agreement that obligates both parties to fulfill the task in a managerial manner. Entities partnering under the PPP umbrella pursue mutually compatible or complementary goals. However, the government retains control over central management responsibilities, whereas the operational responsibility is assigned to the private enterprise (Skelcher 2005). Potential PPPs require carefully performed preevaluations of the project's viability. Typically, a project is financially viable for the private entity if the revenues generated cover its costs and provide sufficient return on investment. For the host government, viability depends on its efficiency compared with the economics of financing the project with public funds, taking into account the expertise that the private entity is expected to bring and the transfer of risk to the private entity.

But besides the many potential advantages, PPPs have many inherent risks (table 5.1). Common risks include (a) asymmetrical dissemination of information, (b) opportunist behavior of private enterprises, (c) lack of transparency, (d) financial risks, (e) uncertainty in task fulfillment, (f) control deficiencies, and (g) a lack of democratic participation and responsibility. Many country-specific

Table 5.1 **Advantages and Risks of Public-Private Partnership Projects**

Advantages	Risks
For public authorities	
Financial relief or gain in efficiency	Selection of partner
More rapid realization of projects	Long-term commitment
Optimized task fulfillment or relief	Complexity
Image enhancement	Conflict of interest or political risks
For the private partner	
Opening of new markets	Long-term commitment and stability
Attractiveness of the public business partner	Long path to decision
Improvement in the chances of success	Pseudocompetition
Anticipation of yields	Diverging interests

Source: Lienhard 2006.

legal issues also exist (such as those related to laws governing competition, awarding of contracts, taxation, employment, and contracts). Hence, carefully structured contracts between the government and private enterprise are necessary (Lienhard 2006). Moreover, the responsible actors require a new culture—one of trust, mutual understanding, and learning from each other—that is aimed at a common public welfare-oriented goal, not simply an economic one. Special challenges exist in including private enterprises in fulfilling sovereign tasks. For NMHSs, weather warnings would be a sovereign task, unlikely to be viewed as appropriate for a PPP. But a PPP could be involved in how warnings are delivered. Thus, a PPP must be based on a win-win situation with the advantages and risks to all of the parties carefully weighed.

PPPs are designed in various ways, which are influenced by political and cultural contexts. In Scandinavia and the United Kingdom, PPPs somewhat contradict the tradition of public good provision by the social democratic welfare state. But such is not the case in the United States, which historically has seen the private sector more closely involved in providing public goods (GAO 1999; Skelcher 2005). In Switzerland, the MeteoSwiss PPP example shows how NMHSs can work with private insurance companies (see box 5.10 technical insight).

How do PPPs mesh with the possible operating models for NMHSs? Generally, all entities within models I to IV can cooperate under a PPP with a private actor (model V). Nevertheless, there may be external effects, such as the legislative or political framework,[7] that may force or hinder establishing PPPs (Gawel 2011).

A clearly defined legal environment is critical because it serves as a legal basis for the government entities to enter into a long-term partnership with private entities. That said, it may affect government entities of models I to IV to a larger extent than private entities. Government entities fundamentally depend on the legal authorization to be active in the PPP field, whereas private actors are free with respect to potential collaborations (ADB 2008; Gawel 2011).

Entities in models I and III and their decisions regarding PPPs are more likely to be affected by the political framework, given that these entities are under direct political control or work at arm's length from the government. In contrast,

Box 5.10 Technical Insight: Public-Private Partnership between MeteoSwiss and Mobiliar

Switzerland hosts many leading insurance and reinsurance companies. Several of them have realized the importance of meteorological and climatological data for their business—for example, property damage through severe weather events or climate-related exposure changes. For several years, MeteoSwiss has been cooperating with some of these companies. Usually, cooperation takes the form of a classic client–service provider relationship, where MeteoSwiss delivers highly specific weather or climate data. However, with some insurance companies, the cooperation goes further. In one case, MeteoSwiss produced a film and a brochure financed by the insurance pool for aircrafts to inform pilots of meteorological hazards. The brochure has become MeteoSwiss's most requested information and is still jointly offered. Another case is the development of seasonal forecasts supported by a major reinsurance company.

Recently, MeteoSwiss and Mobiliar, a Swiss insurance company, have made efforts to establish long-term cooperation. The insurance company hopes to expand client retention and reduce damage costs by improving the flow of weather information to its customers.

In the first phase, a warning system was developed for dangerous weather phenomena, which is now being distributed through a mobile application that is integrated in the company's existing app service. A third partner of the insurance company does the technical implementation, while MeteoSwiss provides the warning logic and data delivery. In addition, MeteoSwiss has an advisory and supporting role. Such a PPP not only generates additional income for MeteoSwiss but also offers the opportunity to share technical knowledge. Synergies can be exploited in such a way to offer joint products and reduce development costs.

In the second phase, the collaboration on customer-specific services will involve a PPP-type approach. MeteoSwiss and Mobiliar will jointly develop additional features for the mobile application—such as hail forecasts for wine growers or outdoor GPS (global positioning system)–based road weather forecasts for motorists—which will be marketed later individually.

Such a PPP can add value for all stakeholders. For Mobiliar, it has the following advantages:

- Positive service differentiation
- Improved image
- Expanded market presence
- Increased client retention
- New customers
- Reduced property damage through better-informed customers

Mobiliar's insurance customers also benefit from the arrangement. They receive

- Firsthand information on hazardous weather phenomena
- Reduced property damage
- Customized severe weather information
- Simple, intuitive, and *state-of-the-art* distribution

box continues next page

Box 5.10 Technical Insight: Public-Private Partnership between MeteoSwiss and Mobiliar *(continued)*

Finally, MeteoSwiss realizes the following advantages:

- Additional long-term revenue (data and advice)
- Joint product development and know-how sharing
- Lower development costs
- Expanded public service

SOEs (model IV) are probably less affected because they typically have more distance from the political tug-of-war.

PPPs are often complex constructs and are highly demanding for the private actor regarding the output provision. Hence, for some project types, the number of potential private partners is limited, an arrangement that gives the private company a quasi-monopoly and thus can greatly influence the negotiation process (ADB 2008). Examples are international vaccination programs, where the potential private partners in the pharmaceutical industry are limited (Buse and Walt 2002).

Legal and Regulatory Frameworks

Throughout the book, we have referred to the importance of establishing the appropriate legal and regulatory framework for the effective operations of NMHSs and delivery of their services. Surprisingly, many countries have no distinct legal framework for meteorological and hydrological forecasts and warnings.

Legal Framework

By international agreement, WMO Members have designated their NMHSs as the authority for producing and delivering meteorological and hydrological warnings. In a properly managed system, a warning triggers actions from other sectors. For example, a warning of a flood might trigger the public health agency to warn of a waterborne disease risk; a storm surge warning should trigger a warning from civil protection authorities to evacuate low-lying areas that are at risk of inundation and should alert local governments to act. The early-alerting action in the Shanghai Multihazard Early-Warning System ensures that the appropriate agencies are notified before any public announcement is made and that the public messages are coordinated. However, such a system should not in any way inhibit issuing meteorological and hydrological warnings, because in many instances, although these warnings do not trigger a widespread emergency response, they are needed to ensure that individuals take appropriate action to protect themselves.

In the absence of a national legal framework, however, NMHSs may have difficulty asserting this authority—and forecasters may be held accountable for forecast errors that are outside the bounds of predictive skills. Contrary to international agreement, in some countries agencies other than the NMHSs may issue weather and water warnings. Or such warnings may be ignored as having no authority, may be contradicted with impunity by commercial weather services, or may not even be issued at all. Actions are not taken that could have protected civil society, and disasters that could have been prevented happen anyway.

Although meteorological laws are generally country specific, a number of commonalities and best practices can be identified by critiquing a few:[8]

- In 1955, Australia was one of the first countries to enact a meteorological law. The law governed the Bureau of Meteorology (BOM), outlining its functions, powers, charges, and regulations. The BOM had the additional responsibility of collecting and analyzing hydrological data through the 2007 Water Act.

- In the United States, the 2005 National Weather Services Duties Act states that the NWS's responsibility is to protect life and property by (a) preparing and issuing severe weather forecasts and warnings designed to protect the lives and property of the general public; (b) preparing and issuing hydrometeorological guidance and core forecast information; (c) collecting and exchanging meteorological, hydrological, climate, and oceanographic data and information; and (d) providing reports, forecasts, warnings, and other advice to the secretary of transportation. The law explicitly forbids the NWS from competing with the private sector.

- In the United Kingdom, the Met Office's activities are defined in the Government Trading Funds Act 1973 as amended and the Meteorological Office Trading Fund Order 1996 and subsequent amendments.[9] In contrast to U.S. law, these acts are explicit about the NMS's achieving a return on investment through commercial services. A similar legal framework exists in Tanzania (Gray 2012). Other countries have also focused on revenue. In Nepal, the government requires all public agencies to sell data and services among the government departments, whereas the Ministry of Finance retains all of the revenue.

- In China, the Meteorology Law states the China Meteorological Administration's responsibilities. The State Council has also issued regulations for meteorological hazards and warnings. Each agency knows its role and responsibilities at all levels of government, which helps ensure that proper actions are taken in response to a warning. Punitive provisions are included in these regulations for people who issue severe-weather alerts in violation of the law (as is done in the Republic of Korea).

- In the Republic of Korea, the Weather Act also defines the responsibilities of the Korea Aviation Meteorological Agency, a subsidiary of the wholly publicly funded Korea Meteorological Administration. The act states that the agency "may engage in any profitable business related to aviation meteorological services, as prescribed by Presidential Decree." The Weather Industry Promotion Act[10] of Korea gives the Korea Meteorological Administration the responsibility to actively assist in developing and enhancing Korea's commercial weather enterprise.

Despite the different political systems underpinning these laws and regulations, the commonalities and best practices form the basic elements of legislation that provide the foundation for operating NMHSs (see box 5.11 technical insight).

Box 5.11 Technical Insight: Critical Elements of a Meteorological or Hydrological Law

An effective meteorological or hydrological law should contain the following elements:

- *Mandate.* The law should be clear about the point of accountability in government and the purpose of the National Meteorological and Hydrological Services (NMHSs) purpose, such as (a) ensuring the accurate and timely issue of forecasts; (b) preventing meteorological and hydrological disasters; (c) properly exploiting and effectively protecting climate resources; and (d) providing services for economic development, social development, and people's well-being.
- *Inclusion of all actors.* The law should include all entities engaged in observation, forecasting, services, disaster prevention, and exploitation of climate resources and research in relevant sciences and technology.
- *Public service.* The law should identify NMHSs as basic public services for economic and social development and people's well-being, with public meteorological and hydrological services being the first priority. NMHSs should be explicitly incorporated into national economic and social development plans and fiscal budgets.
- *Conditions for free services.* The law should state the conditions for free public meteorological services, which should be guaranteed, and should provide for paid services if the public service conditions are satisfied. The public services should extend to key production sectors. For example, in most developing countries, the agricultural sector should receive information to support smallholder farmers.
- *Authority.* The NMHSs should be designated explicitly as the competent authority responsible for meteorological and hydrological work nationally and at lower levels of government as appropriate.
- *Research and development.* The government should (a) encourage and support meteorological research and development, (b) ensure training in the use of advanced meteorological science and technology, (c) protect achievements in meteorological modernization, (d) strengthen international cooperation and exchange, and (e) encourage the development of information technology to improve dissemination of information.
- *External entities.* The law should provide for meteorological and hydrological activities carried out by other countries' organizations and individuals within the national territory to ensure the approval of the NMHSs and to prevent the undermining of public services.
- *Facilities.* The law should provide for constructing and maintaining meteorological facilities, and the state should protect those facilities to ensure normal operation. Equipment standards should be explicit so that instruments are properly verified, maintained, and calibrated.

box continues next page

Box 5.11 Technical Insight: Critical Elements of a Meteorological or Hydrological Law *(continued)*

- *Observations.* The law should include provisions on observations, including the requirement that observations to be maintained by entities other than the NMHSs on their behalf (such as ships, aircraft, hydroplants, and cell phone towers). The relevant environs should be protected to ensure unobstructed observations over time.
- *Forecasts and warnings.* The law should provide for forecasting and warnings, and there should be a unified system for issuing them. Only the competent authority—the National Meteorological Service (NMS) or national hydrological service (NHS)—or its subordinate organizations should issue authoritative meteorological and hydrological forecasts and severe weather and water warnings. Specialized forecasts should be issued for government departments. Radio and television stations and designated newspapers should allocate particular time slots or space every day for these forecasts or severe warnings. The NMS should guarantee the quality of the meteorological forecast programs it has prepared. Information technology departments should ensure unblocked meteorological telecommunication.
- *Radio frequencies.* Wireless frequencies and channels designated for meteorological use should be protected from interference.
- *Revenue generation.* The law should be specific about revenue generation, including cost recovery and charges, commercial services, and the retention of revenue generated. It should consider whether revenue generated on the basis of the NMHSs' information (such as by media that distribute meteorological information) should be used or shared to develop the services.
- *Meteorological and hydrological disasters.* The government should commit to better monitoring and warning systems for meteorological and hydrological disasters, arrange for departments to create plans to prevent them, and take effective measures to increase the capability of preventing them. NMHSs should be authorized to arrange for joint monitoring and forecasting of these events among government departments. NMHSs should also propose timely measures for preventing and assessing the events.
- *Climate service.* The law should provide for climate services and the roles and responsibilities of the NMS and other bodies. The NMS should be responsible for coordinating climate monitoring, analysis, and assessment and for monitoring atmospheric composition that may cause climate deterioration. It should coordinate and arrange for climate feasibility studies related to city planning, key national construction projects, major regional economic development projects, and large projects for exploiting climate resources (such as solar and wind energy).
- *Legal responsibilities.* The law should provide for legal responsibilities—including for the unauthorized issuance of public meteorological forecasts or severe weather warnings, failure to distribute or use the latest authoritative information, and negligence on the part of the staff members responsible for issuing forecasts and warnings.
- *International treaties and cooperation.* The law should conform to international treaties on meteorological activities and should explicitly identify the role of NMHSs in international cooperation.

Regulatory Framework

Countries also need to adopt government regulations to facilitate the responsibilities and authorities under the legislation—for example, preventing and being prepared for meteorological and hydrological disasters. Such a regulatory framework is created to strengthen preparedness for disasters, to prevent and reduce losses, and to protect lives and property. It spells out the responsibilities of the NMS and NHS at all levels of government (from the national to the local level) and identifies the responsibilities of the national and local governments. It may also stipulate how certain public institutions, such as schools, should be prepared and how individuals and organizations can become involved in the countrywide effort.

In some cases, regulations may explicitly give the lead to the NMS to develop plans for the prevention of and preparedness for meteorological disasters. Such regulations anticipate that government institutions will respond appropriately to strengthen the construction of facilities to prevent meteorological disasters and expect all weather-sensitive sectors to account for meteorological and hydrological hazards in establishing construction standards.

The NMS may also lead the development of an emergency response plan, or it may work closely with a civil protection agency to develop such a plan. It is frequently observed that the NMHSs do not figure prominently in the development of some national emergency preparedness plans and that there is no effective operational working relationship between civil protection and the NMHSs. A regulatory framework is designed to identify the roles and responsibilities of each of the stakeholders and to support a more effective hazards partnership.

The regulations should also clearly state how warning information is disseminated and used, with the expectation that government departments will follow an approved course of action and the media will communicate the information in a timely and correct fashion to the public, ensuring that the authoritative nature of the warning is preserved. Regulations may also provide for how information is broadcast to the public, using all available media, including transportation networks and public facilities.

Finally, the regulations should consider the legal liabilities associated with (a) failing to formulate plans for meteorological disaster prevention, (b) failing to take measures to prevent disasters, (c) issuing qualification certificates to unqualified installations, (d) concealing warnings or issuing false warnings, and (e) failing to take timely emergency response measures.

International Agreements

A number of international agreements oblige governments through government agencies, such as NMSs. These agreements include the Convention of the World Meteorological Organization,[11] which states the purpose of the WMO and ways NMSs contribute to this purpose—in particular, that all Members must do their utmost to implement the decisions of the World Meteorological Congress.

Additional agreements include the International Convention for the Safety of Life at Sea,[12] which governs the provision of marine forecasts; annex 3 of the Convention on International Civil Aviation,[13] which governs the provision of meteorological services for international air navigation; and the United Nations Framework Convention on Climate Change, which reflects the role some NMSs play in national adaptation programs of action.[14]

So far, this report has provided the basics about the purpose, organization, and operation of NMHSs, along with a look at their current strengths and weaknesses in practice. The next chapter provides guidance on how to modernize NMHSs—and how the World Bank can best help these institutions, working in collaboration with the WMO and other development partners.

Notes

1. An operating model describes the rationale of how an organization creates, delivers, and captures economic, social, or other forms of value. In theory and practice, the term *operating model* is used for a broad range of informal and formal descriptions to represent core aspects of a business, including purpose, offerings, strategies, infrastructure, organizational structure, trading practices, and operational processes and policies.
2. There are 177 hydrological advisers nominated by members of the World Meteorological Organization (WMO). According to the WMO's general regulations, they should be representatives of the National Hydrological Service.
3. Others include Water and Sewerage Corporation in The Bahamas; the National Water Agency in Brazil; the Finnish Environment Institute; the Hydrological Service, Ministry of Energy, in Iceland; the Department of Irrigation and Drainage in Malaysia; the Ministry of Land, Transport, and Maritime Affairs in the Republic of Korea; the Federal Office for Water and Geology, Swiss Hydrological Survey; the Ministry of Lands, Survey, and Natural Resources in Tonga; and the Centre for Ecology and Hydrology in the United Kingdom.
4. In Scotland, the Flood Forecasting Centre is a joint venture between the Scottish Environmental Protection Agency and the Met Office.
5. Ferlie, Lynn, and Pollitt (2005) published a handbook of public management covering theoretical aspects, such as basic frameworks and disciplinary perspectives, as well as national and international comparisons. Verhoest et al. (2012) provide a comprehensive overview of the development, characteristics, and differences of government agencies in 30 countries, including developing countries. Pollitt et al. (2005) provide a comparison of the institutional pattern and organization of NMHSs in Finland, the Netherlands, Sweden, and the United Kingdom.
6. These criteria were developed and kindly provided by Andreas Lienhard and Etienne Huber of the Center of Competence for Public Management at the University of Bern.
7. The *political framework* refers to the formal legal institutions of a state and describes the behavior within the system and the functioning of the state.
8. For examples of such laws, see Australia's Meteorology Act 1955 (http://www.comlaw.gov.au/Details/C2007C00548), China's Meteorology Law 1999 (http://www.china.org.cn/environment/2007-08/27/content_1034467.htm), Germany's Law on the Deutscher Wetterdienst 1998 (http://www.dwd.de/bvbw/generator/DWDWWW/

Content/Oeffentlichkeit/KU/KUPK/Wir__ueber__uns/Gesetz__PDF__en,templateId =raw,property=publicationFile.pdf/Gesetz_PDF_en.pdf), the Republic of Korea's Weather Act 2009 (http://web.kma.go.kr/eng/images/down/weather_act.pdf), New Zealand's Meteorological Services Act 1990 (http://www.legislation.govt.nz/act/public/1990/0100/latest/whole.html), South Africa's Weather Service Act 2001 (http://metzone.weathersa.co.za/images/pdf_docs/SAWS_act_8_of_2001.pdf), and the U.S. National Weather Services Duties Act 2005 (http://thomas.loc.gov/cgi-bin/query/z?c109:s786:).

9. The financial expectations of the government for the Met Office are laid out in a treasury minute, which states the desired financial objectives of the Met Office Trading Fund in the form of a surplus on ordinary activities before interest (payable and receivable) and dividends expressed as a percentage of average capital employed. Thus, in addition to covering costs, the Met Office is expected to provide a return on the government's investment.
10. The text of the act is available at http://web.kma.go.kr/eng/images/down/wip_act.pdf.
11. The text of the convention is available at http://www.eco.gov.az/hidromet/umumdunya%20meteorologiya.pdf.
12. The text of this convention is available from the International Maritime Organization at http://www.imo.org/About/Conventions/listofconventions/pages/international-convention-for-the-safety-of-life-at-sea-%28solas%29,-1974.aspx.
13. The text of this convention is available through the Swiss government website (http://www.bazl.admin.ch/dokumentation/grundlagen/02643/index.html?lang=de), along with other International Civil Aviation Organization documents.
14. An example of such a program is the Climate Change National Adaptation Programme of Action of Ethiopia, in which the Ethiopian National Meteorological Agency plays the leading role (Tadege 2007).

References

ADB (Asian Development Bank). 2008. *Public-Private Partnership Handbook*. Manila: ADB. http://beta.adb.org/documents/public-private-partnership-ppp-handbook.

Bade, Stephanie, Stephan von Grünigen, and Walter Ott. 2011. *Der volkswirtschaftliche Nutzen von Meteorologie in der Schweiz: Verkehr und Energie* [The Economic Value of Meteorology for Switzerland: Traffic and Energy]. Zurich, Switzerland: Econcept/MeteoSwiss. http://www.meteoschweiz.admin.ch/web/de/meteoschweiz/medien/medienmitteilungen/wert_meteo.Related.0002.DownloadFile.tmp/vonumet.pdf.

Broussine, Mike. 2009. "Public Leadership." In *Public Management and Governance*, edited by Tony Bovaird and Elke Löffler, 261–77. New York: Routledge.

Buse, Kent, and Gill Walt. 2002. "The World Health Organization and Global Public-Private Health Partnerships: In Search of 'Good' Global Governance." In *Public-Private Partnerships for Public Health*, edited by Michael R. Reich, 169–95. Cambridge, MA: Harvard Center for Population and Development Studies.

Ferlie, Ewan, Laurence E. Lynn, and Christopher Pollitt, eds. 2005. *The Oxford Handbook of Public Management*. New York: Oxford University Press.

Freebairn, John W., and John W. Zillman. 2002. "Funding Meteorological Services." *Meteorological Applications* 9 (1): 45–54.

Frei, Thomas. 2009. "Economic and Social Benefits of Meteorology and Climatology in Switzerland." *Meteorological Applications* 17 (1): 39–44.

GAO (U.S. General Accounting Office). 1999. "Public-Private Partnerships: Terms Related to Buildings and Facility Partnerships." GAO, Washington, DC. http://www.gao.gov/special.pubs/Gg99071.pdf.

Gawel, Erik. 2011. "Political Drivers of and Barriers to Public-Private Partnerships: The Role of Political Involvement." Working Paper 98, Faculty of Economics and Management Science, Leipzig University, Leipzig, Germany.

Gill, Derek. 2002. "Signposting the Zoo: From Agencification to a More Principled Choice of Government Organisational Forms." *OECD Journal on Budgeting* 2 (1): 27–79.

Gray, Mike. 2012. "NMS Data Availability Study for World Bank." Met Office, Exeter, U.K.

Greve, Carsten, Matthew Flinders, and Sandra Van Thiel. 1999. "Quangos: What's in a Name? Defining Quangos from a Comparative Perspective." *Governance: An International Journal of Policy and Administration* 12 (2): 129–46.

Hallegatte, Stephane. 2012. "A Cost Effective Solution to Reduce Disaster Losses in Developing Countries: Hydrometeorological Services, Early Warning, and Evacuation." Policy Research Working Paper 6058, World Bank, Washington, DC.

Hughes, Owen E. 2003. *Public Management and Administration*. New York: Palgrave Macmillan.

Jean, Michel, Bruce Angle, David Grimes, and John Falkingham. 1999. "Structure and Evolution of National Meteorological and Hydrological Services: International Comparisons." *WMO Bulletin* 48 (2): 159–65.

Lazo, Jeffrey K., Rebecca E. Morss, and Julie L. Demuth. 2009. "300 Billion Served." *American Meteorological Society* 90 (6): 785–98.

Lienhard, Andreas. 2006. "Public Private Partnerships (PPPs) in Switzerland: Experiences, Risks and Potentials." *International Review of Administrative Sciences* 72 (4): 547–63.

Met Office. 2007. "Met Office Framework Document." Met Office, Exeter, U.K. http://www.metoffice.gov.uk/media/pdf/6/4/Met_Office_Framework_document.pdf.

NRC (National Research Council). 2003. *Fair Weather: Effective Partnerships in Weather and Climate Services*. Washington, DC: National Academies Press.

———. 2012. *Weather Services for the Nation: Becoming Second to None*. Washington, DC: National Academies Press.

OECD (Organisation for Economic Co-operation and Development). 1995. "Recent Trends in Privatisation." OECD, Paris. http://www.oecd.org/dataoecd/57/35/1898010.pdf.

Pollitt, Christopher, Colin Talbot, Janice Caulfield, and Amanda Smullen. 2005. *Agencies: How Governments Do Things through Semi-Autonomous Organizations*. New York: Palgrave Macmillan.

SFC (Swiss Federal Council). 2006. *Bericht des Bundesrates zur Auslagerung und Steuerung von Bundesaufgaben (Corporate-Governance-Bericht)*. Corporate governance report, Bern. http://www.efd.admin.ch/dokumentation/zahlen/00578/01061/index.html-?lang=de&download=NHzLpZeg7t,lnp6I0NTU04212Z6ln1acy4Zn4Z2qZpnO2Yuq2Z6gpJCDdIR_e2ym162epYbg2c_JjKbNoKSn6A.

Skelcher, Chris. 2005. "Public-Private Partnerships and Hybridity." In *The Oxford Handbook of Public Management*, edited by Ewan Ferlie, Laurence E. Lynn, and Christopher Pollitt, 347–70. New York: Oxford University Press.

Steiner, Reto, and Etienne Huber. 2012. "Agencification in Continental Countries: Switzerland." In *Government Agencies: Practices and Lessons from 30 Countries*, edited by Koen Verhoest, Sandra Van Thiel, Geert Bouckaert, and Per Lægreid, 191–202. New York: Palgrave Macmillan.

Steiner, Thomas J., John. R. Martin, Neil D. Gordon, and Malcolm A. Grant. 1997. "Commercialisation in the Provision of Meteorological Services in New Zealand." *Meteorological Application* 4 (3): 247–57.

Tadege, Abebe. 2007. *Climate Change National Adaptation Programme of Action of Ethiopia*. National Meteorological Agency, Addis Ababa. http://unfccc.int/resource/docs/napa/eth01.pdf.

UNCTAD (United Nations Conference on Trade and Development). 2011. *World Investment Report 2011*. Geneva, Switzerland: UNCTAD. http://www.unctad-docs.org/files/UNCTAD-WIR2011-Full-en.pdf.

Van Thiel, Sandra. 2001. *Quangos: Trends, Causes, and Consequences*. Farnham, U.K.: Ashgate.

———. 2009. "The Rise of the Executive Agencies: Comparing the Agencification of 25 Tasks in 21 Countries." Paper presented at the annual conference of the European Group for Public Administration, St. Julian's, Malta, September 2–5.

———. 2012. "Comparing Agencies across Countries." In *Government Agencies: Practices and Lessons from 30 Countries*, edited by Koen Verhoest, Sandra Van Thiel, Geert Bouckaert, and Per Lægreid, 18–26. New York: Palgrave Macmillan.

Verhoest, Koen, Sandra Van Thiel, Geert Bouckaert, and Per Lægreid, eds. 2012. *Government Agencies: Practices and Lessons from 30 Countries*. New York: Palgrave Macmillan.

Weiss, Peter. 2002. "Borders in Space: Conflicting Public Sector Information Policies and Their Economic Impacts." Summary report, U.S. National Weather Service, Washington, DC.

WMO (World Meteorological Organization). 2012. "Results of the Survey on the Role and Operation of National Meteorological and Hydrological Services Conducted in June–August 2011." WMO, Geneva, Switzerland.

Zillman, John W. 1999. "The National Meteorological Service." *WMO Bulletin* 48 (2): 129–59.

———. 2003. "The State of National Meteorological Services around the World." *WMO Bulletin* 52 (4): 360–65.

Zillman, John W., and John W. Freebairn. 2001. "Economic Framework for the Provision of Meteorological Services." *WMO Bulletin* 50 (3): 206–15.

CHAPTER 6

Guidance on Modernizing NMHSs

In This Chapter

Modernizing National Meteorological and Hydrological Services (NMHSs) cannot be piecemeal. The process should be transformative, ensuring that NMHSs can deliver the services that their stakeholders expect. Partial modernization of observation networks tends to have little effect if the NMHSs cannot produce better forecasts and deliver more accurate warnings. Improvement of services and modernization of infrastructure come with a price. Ongoing operating and maintenance (O&M) costs of new networks and communication systems—along with the need to attract and retain more qualified personnel to run new networks—add to the costs of doing business for NMHSs, so these factors should be taken into account in the planning stage. The World Bank is scaling up its support to NMHSs' modernization efforts and consolidating the experiences of emerging good practices.

Introduction

For developing and least developed countries, there is no definitive approach to modernizing NMHSs. But over the past decade, the World Bank has gained considerable experience in trying to do so. One of the difficulties is that providing weather and climate services is a complex process that depends on advanced technologies, modern communication systems, and highly skilled employees—all of which are in short supply in many countries. Other difficulties involve inadequate incentives in many countries for public services, along with the reality that the institutions themselves have limited political visibility and often lack modernization champions.

This chapter describes the practical experiences of the World Bank and the Global Facility for Disaster Reduction and Recovery (GFDRR) Hydromet team and makes some recommendations that can guide World Bank teams and other practitioners through this complex process. It also draws on the combined experience of major modernization efforts that have been successfully completed by several advanced NMHSs. Experience shows that it is critical that the modernization process not be done in a one-off, uncoordinated, or piecemeal manner; instead modernization must be approached as a holistic effort.

Modernizing Advanced Meteorological and Hydrological Services

Let us begin with the experiences of industrial countries. Typically, NMHSs have a built-in process to regularly modernize. Modernization is part of the life-cycle management of assets and networks, complemented by specialized programs that introduce new networks (for example, Doppler radar) or technologies. If modernization needs significantly exceed the regular agency funding allocations, special targeted programs are prepared and implemented. Most NMHSs implement new technologies as part of their strategic development. And most maintain research and development (R&D) expertise within their services or work closely with universities or government institutes to test new capabilities before implementation.

But the 20th century saw exponential growth in the technological capabilities of weather observations and forecasting, which made it tough for National Meteorological Services (NMSs) to keep pace. Even in the United States, by the 1980s, a major modernization effort was needed owing to the lack of continuous investment. Between 1989 and 2000, the U.S. government invested US$4.5 billion to implement the modernization and associated restructuring of the National Weather Service (NWS). This effort involved adopting new observational and computational systems, redefining the NWS field offices, and restructuring the workforce.

Yet despite the NWS's highly skilled workforce and technical know-how, major implementation problems arose, along with considerable cost overruns. As might have been expected, the budget, schedule, and technological issues encountered reflected the traditional challenges of large projects: (a) inexperience of the government project-level leadership, (b) shifting budget constraints, (c) ambitious technology leaps, (d) multiparty stakeholder pressures, (e) cultural inertia, (f) contractor shortcomings, and (g) oversight burden.

What lessons can be drawn from the NWS's experience? A study by the National Research Council found that because policy makers allowed the NWS systems to become nearly obsolete, their modernization and restructuring became very complex and expensive, thereby creating problems with the staff and external stakeholders (NRC 2011, 2012). Other findings included the following:

- The time-scale for the government system procurement process is very long compared with that for implementation of major technological changes. The pace of technological progress complicates planning, procurement, and deployment of large, complex systems.
- NMHSs' modernization can face resistance from employees, stakeholders, and sometimes the general public.

- The execution of NMHSs' modernization programs require working with many partners, which may provide cost sharing and improve the understanding of user needs.
- The modernization and associated restructuring of the NWS show that candid yet nonadversarial advice from outside experts and other interested parties is useful in designing and developing a large complex system.

These lessons apply to many other countries equally, particularly those that have allowed their NMSs to decline to the point of obsolescence. Following modernization and to prevent reversion to the same cycle of modernization and obsolescence, the NWS subsequently created a program that focused on continuously infusing new technologies. Working within the National Oceanic and Atmospheric Administration, the NWS is now developing new capabilities within government laboratories or through external R&D sources in response to ongoing requirements for product and service improvement. The NWS also created the post of science operations officer, who serves to rapidly integrate scientific advancements into Weather Forecast Office operations.

At the U.K. Met Office, technical innovation and a continuous infusion of new capabilities have been a hallmark of the office's success, thanks to a very strong internal R&D program focused on solving operational problems, as well as access to high-quality, university-based research. The Australian Bureau of Meteorology, the China Meteorological Administration (CMA), Météo France, the Meteorological Service of Canada, and many other leading NMSs remain at the forefront of technical innovation because of this approach. This approach also provides the means for developing and testing operation prototypes with users and stakeholders to refine system design.

Modernizing NMHSs in Developing Countries

In middle-income countries, there are a few cases of successful modernization program, some of which were undertaken jointly with the World Bank by countries such as Poland, the Russian Federation, and Turkey. The modernization of the CMA is an example of a particularly successful program. It was driven by strong economic arguments on the value and economic benefits of NMSs, which helped build a consensus for support that resulted in massive state funding complemented by funding from municipal governments and fee-based sources (see chapter 5). Since 2000, the CMA budget has been growing faster—at a rate of about 15 percent—than China's gross domestic product. A few other countries (such as Indonesia, Morocco, and Tunisia) have achieved better hydrometeorological service delivery by relying on national and international resources and expertise.

Many Members of the World Meteorological Organization (WMO)—such as China, Finland, France, Japan, the Republic of Korea, Spain, Switzerland, the United Kingdom, and the United States—and donor agencies actively participate in modernizing NMHSs in developing countries. Some of these efforts, such as those in the Caribbean, seem to be operational mostly because of long-term and ongoing donor support. But elsewhere, most of these efforts have proved to be unsustainable for a number of reasons, including (a) underlying weaknesses in national public services resulting from a lack of government commitment, (b) low human capacity, (c) inadequate design of new systems that are ill suited to the operational environment, and (d) inadequate coordination among donors within specific countries and among countries.

Of course, the lessons from these failures should be carefully evaluated and absorbed. But there are few reports to draw on that review the results of donor assistance after project completion. One that does exist is the report on Swiss support to hydrometeorological services in the Aral Sea Basin, which was written during the closing of the eight-year Swiss regional program in Central Asia (Zaitsev and Yagupov 2010). It anticipated a low sustainability of the program results, owing chiefly to lack of government operational funding and lack of trained personnel. Another example would be the broad modernization of Mexico's NMS, initiated in the mid-1990s. Although the proposed modernization was sound—and supported throughout by WMO advice—the infrastructure later deteriorated owing to lack of funds.

In fact, a common reason advanced equipment—such as automatic weather stations, telemetric systems, and Doppler radar—are not long-lived in developing countries is the lack of O&M funds and trained personnel. Consider Mozambique, which received considerable donor support from the European Union, Finland, Spain, and other economies after massive floods in 2000. That support included the installation of modern equipment (such as several dozen automatic weather stations, a hydrological telemetric system, and two Doppler radars) and training. But a few years after their installation, these expensive instruments and systems became inoperable, mostly owing to a lack of basic maintenance (unattended automatic instruments), resources (such as the lack of diesel fuel for uninterrupted power supply for Doppler radar), inadequate design, and vandalism. In some countries (for example, Vietnam), donors are supporting the modernization of separate elements of the hydromet system or the creation of small regional networks within the system without making necessary systemic changes to the entire national system to fully integrate these networks. This lack of integration creates bottlenecks and inefficiencies in operations.

As a result, a recent review of WHYCOS (World Hydrological Cycle Observing System) stresses the need for a holistic approach to strengthen national hydrological services (NHSs) by focusing on sustainable outcomes (such as the provision of flood forecasts and warnings) and not solely on outputs (such as the acquisition and distribution of hydrological data) (Pilon and Asefa 2011). Otherwise, long-term support is difficult to maintain. The report highlights (a) the critical role of the WMO in overseeing the WHYCOS program; (b) the commitment of participating countries to the long-term sustainability through increased operations, maintenance, and recapitalization support; and (c) such countries' commitment to the free and unrestricted exchange of hydrological data (Pilon and Asefa 2011). However, the commitment to data sharing goes beyond Resolution 25 of the 13th World Meteorological Congress in that it implies that the countries will exchange data rather than simply promote the concept. The review also contends that a lack of commitment to the free and unrestricted exchange of data between participating countries is a major limitation to developing effective national hydrological and meteorological services—and that the free and unrestricted exchange of data should be reflected in legal agreements in all donor-supported hydrological and meteorological projects. The emphasis on flood forecasting and warnings also highlights the importance of very close cooperation between NMSs and NHSs.

World Bank Experience in NMHSs Modernization

For the World Bank, modernizing NMHSs is a relatively new area of technical assistance and investment support. Until the mid-1990s, investments in NMHSs were structured as small-scale activities within water resource management, agricultural, or emergency operations—typically motivated by flooding and flood control (Hancock and Tsirkunov 2013). The focus was on fragmented systems (contributing to weather and climate services that are designated to serve a single sectoral user) and partial systems (aiming to improve forecasting for more than one user but not addressing the needs of all users). A systemic problem appears to have been the failure to integrate observations and forecasting into warning systems.

Since the mid-1990s, there has been an overall rising trend in hydromet investments (see figure 6.1) and a gradual shift toward developing more complete systems (upgrades that serve all or many users and implicitly aim at being sustainable). Even so, a recent review by the World Bank Independent Evaluation Group indicates that the sustainability of the hydromet investments remains an issue, especially for Africa, which must cope with a high level of climate variability (Chomitz, Akhmetova, and Hutton 2012). The report notes that in Africa the network of hydromet stations is sparse and deteriorating, hydromet data are often spotty and inaccurate, and existing stations are often not functioning or fail to communicate with the global meteorological network. Only 4 of 12 closed African projects reported attention to maintenance, and only in the Senegal River Basin did the self-evaluation report consider sustainability to be likely (Chomitz, Akhmetova, and Hutton 2012).

Table 6.1 describes some World Bank hydromet programs. Some of the major systems projects, which are now complete, were implemented in Poland (box 6.1), Turkey, and Russia (see box 6.2)—the latter ranking as the World Bank's largest hydromet modernization project at US$173 million (2005–13). In Poland and Turkey, hydromet modernization was triggered by a major disaster that convinced governments of the importance of a well-functioning NMS for early-warning systems and mitigating natural hazard risks.

In 2012, new hydrometeorological modernization projects were initiated for Mozambique, Nepal, Vietnam, the Republic of Yemen, and other countries (table 6.2). In Nepal, the project is one of the few examples of an end-to-end

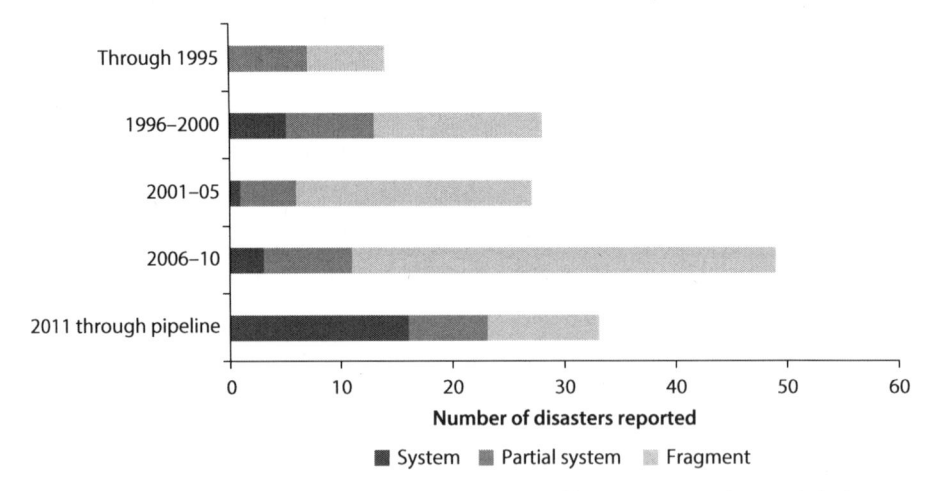

Figure 6.1 World Bank Investments in Hydromet, Fiscal Years 1995–2011

```
Through 1995        ▇▇▇▇▇▇▇
1996–2000           ▇▇▇▇▇▇▇▇▇▇▇▇▇▇
2001–05             ▇▇▇▇▇▇▇▇▇▇▇▇▇
2006–10             ▇▇▇▇▇▇▇▇▇▇▇▇▇▇▇▇▇▇▇▇▇▇▇▇
2011 through pipeline ▇▇▇▇▇▇▇▇▇▇▇▇▇▇▇▇
                    0    10    20    30    40    50    60
                        Number of disasters reported
                ■ System   ■ Partial system   ▒ Fragment
```

Source: Hancock and Tsirkunov 2013. Implementation completion reports of the projects were used to compile the figure.

approach, focusing on institutional strengthening, modernization of the observation and forecasting systems, and service delivery. But so far, the World Bank has limited experience in NMS strengthening in least developed and impoverished countries.

A big problem for these countries is retaining highly skilled employees, such as forecasters and information technology personnel. Competition with the private sector is substantial as economies develop, and the capacity to reward is much higher in the private sector. Innovative ways need to be found for staffing NMHSs in these countries. Such methods may include establishing special, higher-remuneration categories of employment for technical posts in the public sector; using contract personnel; providing incentives through specialized training; using public-private partnerships; and outsourcing. But retaining highly skilled employees will not be easy given government resistance to higher remuneration for those with high qualifications; often technical workers are paid at the same level as unskilled workers. A recent exception is Cambodia: in the Ministry of Water Resources and Meteorology, employees who operate the new radar and forecasting offices in the Department of Meteorology are paid at a higher rate, although remuneration is made through a special provisional arrangement supported by the minister.

What lessons can be learned from the World Bank's experience so far with modernizing NMHSs? The key lessons, as box 6.3 highlights, are (a) taking an inclusive end-to-end approach that is transformative; (b) giving NMHSs the authority to provide warnings; (c) wherever possible, not waiting until the NMHSs are obsolete to undertake modernization, because this approach makes the job more costly and complex; and (d) engaging staff and external stakeholders to provide better public weather, climate, and water services.

Table 6.1 World Bank Hydromet Programs

Project or component title	Funding (US$)	Description	Status (as of November 2012)
Mexico: Water Resources Management Project (1996–2002)	41.0 million	Overall modernization of meteorology and hydrology	Complete
Poland: Emergency Flood Recovery Project (1997–2006)	62.0 million	Observation networks, forecasting, service improvement	Complete
Peru: El Niño Emergency Assistance (1997–2002)	7.0 million	Forecasting	Complete
Dominican Republic: Hurricane Georges Emergency Recovery Project (1998–2002)	10.0 million	Observation networks and forecasting	Complete
Turkey: Emergency Flood and Earthquake Recovery Project (1998–2005)	26.0 million	Observation networks and forecasting	Complete
Russian Federation: Hydromet Modernization Project I (2005–13)	173.0 million	Institutional strengthening, observation networks forecasting, and service delivery	Complete, implementation completion report under preparation
Sri Lanka: Dam Safety and Water Resources Planning (2008–12)	8.5 million	Observation networks (mostly hydrological) and forecasting	Under implementation
Albania: Disaster Risk Mitigation and Adaption Project (2008–12), Component 2	1.8 million	Strengthening of hydrometeorological services	Under implementation
Moldova: Disaster and Climate Risk Management Project (2010–15)	9.0 million	Observation networks, forecasting, and agrometeorology	Under implementation
India: Bihar Koshi Flood Recovery Project (2010–15), Component 3	30.0 million	Observation networks and flood forecasting	Under implementation
Afghanistan: Irrigation Restoration and Development Project (2011–16)	5.0 million	Establish hydromet service	Under implementation
Central Asia: Hydrometeorology Modernization Project (2011–16)	27.7 million	Institutional strengthening, observation networks and forecasting, service delivery	Under implementation
Mexico: Modernizing the National Meteorological Service to Address Variability and Climate Change in the Water Sector in Mexico (2012–17)	105.0 million	Institutional strengthening, observation networks, forecasting, and service delivery	Under implementation

---

Box 6.1 Poland's National Meteorological Service Overhaul

In 1997, Poland's worst flood resulted in 55 deaths and the loss of US$3.4 billion, or about 2.4 percent of the country's gross domestic product. In response, the government asked the World Bank to prepare the Poland Emergency Flood Recovery Project—which included a US$62 million monitoring, forecasting, and warning system component, implemented by the Institute of Meteorology and Water Management, Poland's national meteorological service (NMS). The main investments included (a) installing about 1,000 meteorological and

box continues next page

Box 6.1 Poland's National Meteorological Service Overhaul *(continued)*

hydrological measuring stations, (b) procuring a supercomputer and developing a computing system for a high-resolution weather forecasting model, (c) upgrading data processing and transmission systems, (d) introducing a lightning detection system, (e) enhancing the weather radar system with eight Doppler radars, and (f) developing a hardware and software platform to provide users with access to forecasts by telephone or the Internet (client service system).

As a result, the quality of services has improved, as shown by better performance indicators and client satisfaction ratings. But the World Bank's implementation completion report also cites several shortfalls: (a) the late introduction of the integrator in the project, (b) a lack of government commitment to increase the NMS budget in line with operation and maintenance needs of a new system, and (c) difficulties in retaining qualified staff members.

Box 6.2 The Russian Federation's Roshydromet Modernization

In the mid-1980s, the capacity of Roshydromet to provide services to Russia and globally was steadily declining, mostly owing to inadequate funding, which was below 50 percent of the basic operational needs. The decline affected all elements of the system: (a) observation networks; (b) data transmission, archiving, and processing facilities; (c) forecasting; (d) research and development facilities; and (e) workforce quality. After more than 20 years of decline, the government recognized the challenge of modernizing such a complex hydromet system and requested World Bank assistance in project preparation and implementation. This request was processed as the first self-standing and integrated hydromet modernization project in the World Bank's portfolio.

The project's objective was to increase the accuracy of forecasts provided to the Russian people and economy by modernizing key elements of the technical base and strengthening Roshydromet's institutional arrangements. The project began in 2005 and was finished in 2013. At US$173 million, it was the World Bank's largest hydromet modernization project. The project was a success, reaching or exceeding the main agreed performance indicators:

- Increased lead time and accuracy of global and regional forecasts
- Improved data collection and transmission
- Drastic reduction of response time for requests of archived data
- Increased reliability of seasonal flow forecasts in the pilot river basins
- Introduction of client satisfaction surveys in Roshydromet performance evaluation
- Development and government approval of a long-term Roshydromet strategy, which led to a massive increase in government support to the agency—from US$76.6 million in 2003 to US$570 million in 2011.

Because the first modernization project could not address all problems accumulated during more than 20 years of underfunding, the second stage of the Roshydromet modernization program is under preparation.

Table 6.2 World Bank NMHSs' Modernization Projects in Poorer Countries

Project or component title	Funding (US$)	Funding source	Stage of preparation (as of November 2012)
Vietnam: Managing Natural Hazards Project—Component 2, Strengthening Weather Forecasting and Early Warning	30.0 million	International Development Association	Approval by the Board of Directors
Lao PDR: Adaptable Program Loan Mekong Integrated Water Resources Management Project—Subcomponent 2.4, Strengthening of Department of Meteorology and Hydrology	2.0 million	International Development Association	Approval by the Board of Directors
Nepal: Building Resilience to Climate-Related Hazards	31.0 million	PPCR and International Development Association	Appraisal
Russian Federation: Hydromet Modernization Project II	141.5 million	International Bank for Reconstruction and Development	Appraisal
Mozambique: Strengthening Hydrological and Meteorological Information Services for Climate Resilience	15.0 million	PPCR and International Development Association	Preparation
Yemen, Rep.: Climate Information System and PPCR Coordination	19.0 million	PPCR	Preparation
Zambia: Strengthening Climate Resilience (PPCR Phase II)	5.0 million	PPCR	Preparation
Cambodia: Adaptable Program Loan Mekong Integrated Water Resources Management Project Component	5.0–6.0 million	International Development Association	Identification
Ghana: Strengthening Hydrological and Meteorological Agencies	15.0–25.0 million	International Bank for Reconstruction and Development and International Development Association	Identification
Africa: Climate Risk Management Project	25.0 million	International Development Association	Identification

Note: PPCR = Pilot Program for Climate Resilience.

Recommendations for Designing and Implementing Modernization Projects

In light of these lessons, how should governments proceed in modernizing their NMSs? No definitive approach exists to designing these projects in developing and least developed countries because of many various elements—such as institutional setups, NMS capacity, natural conditions, and project objectives. At the same time, we believe that integrated or systemic hydromet modernization should include the following groups of interlinked and complementary activities or components:

1. Institutional strengthening, capacity building, and implementation support
2. Modernization of observation infrastructure and forecasting
3. Enhancement of the service delivery system

These three components are discussed in box technical insights 6.4, 6.5, and 6.6.

Box 6.3 Lessons Learned from the Hydromet Modernization Process

1. A systematic, end-to-end approach to modernization that is integrated through all National Meteorological and Hydrological Services (NMHSs) is highly recommended.
    - Take an integrated agencywide approach. Such an approach is more likely to engage the government in focusing on sustainability, which depends critically on government investment beyond external donor support to cover operating and maintaining the new systems and services and retaining a qualified staff.
    - Avoid investing in isolated elements of the National Meteorological Service (NMS) system because such efforts are likely to be unsustainable. The effort needs to be sufficient to enable the transformation of NMSs from largely data collection agencies to full public services. Building institutional capacity is critical for the sustainability of the modernization effort. This approach is far more challenging than building infrastructure.
    - Recognize that no universal or quick solution exists for improving NMHSs. Long-term support to NMHSs is needed (10 years or more), but currently long-term financial instruments are lacking.
    - Encourage better coordination among donors and engage the programs and Members of the World Meteorological Organization (WMO) at an early stage.
2. The NMS must have the authority to provide meteorological warnings.
    - If necessary, enact legislation that creates a meteorological law and regulations that explicitly recognize the NMS and National Hydrological Service as the authoritative voices for meteorological and hydrological warnings and that provide a foundation for all other activities.
3. If NMHSs wait until they are nearing obsolescence, complex and very expensive programs will be required to modernize—as was the case with the U.S. National Weather Service and most World Bank–supported modernization projects, such as Roshydromet.
    - Act in a timely manner is important.
    - Encourage governments to support a continuous process of technological infusion following the initial modernization program to avoid falling behind the evolving societal need for weather, climate, and hydrological information and warnings.
    - Recognize the importance of access to research and development, and retaining and upgrading staff technical skills.
4. External advice from outside experts is important in designing and developing a large complex system, but so is engaging staff members whose careers and livelihoods will be affected by the changes.

As for implementing hydromet modernization projects or components, we would offer the following recommendations drawing on the World Bank's completed projects in Poland, Russia, and Turkey, as well as emerging experience from ongoing operations. These recommendations center on core elements in the project preparation phase, although the detailed activities will be determined by country specifics, the NMHSs' capacity to manage change, available resources, and access to skills in neighboring countries.

> **Box 6.4 Technical Insight: Component 1: Institutional Strengthening, Capacity Building, and Implementation Support**
>
> Component 1 aims to (a) strengthen the legal and regulatory framework of the National Meteorological Service (NMS); (b) improve its institutional performance as the main provider of weather, climate, and hydrological information for the country; (c) build capacity of its personnel and management; (d) ensure operability of future networks; and (e) support project implementation. The approach applies to both NMSs and National Hydrological Services (NHS). Where they are separate organizations, ensuring close collaboration is vital to enable flood forecasting and management.
>
> The institutional component can have three groups of activities or subcomponents:
> - Institutional strengthening, which may include such activities as
>   - Review of institutional development and networks restructuring options
>   - Development of a legal and regulatory framework for the NMS operations, including assessing new business models and enhancing public-private partnerships
>   - Twinning support between the NMS and one or more advanced NMSs[a]
> - Capacity building, which may include
>   - NMS staff training, retraining, and professional development
>   - Professional orientation of the NMS senior management
>   - Education support for the staff
>   - Joint workshops with the major users of the NMS's products and services, including agriculture, emergency, health, water resource management, energy, and transportation
> - Systems design and implementation support, which may include
>   - Detailed design of the NMS systems and implementation support through general consultant or systems integrator or similar technical support
>   - Development needs assessment and design of the observation and monitoring networks
>   - Project management, monitoring, and evaluation
>
> a. Pairing or twinning an NMS in a developing or least developed country with an NMS in a more advanced country is a way to ensure sustainability of the service provision, especially if staffing is weak in the developing country's NMS, which is often the case. The advanced twin agrees to provide technical expertise and know-how, including numerical weather prediction (NWPs) and other guidance that the beneficiary can use in its service provision.

Identifying Projects

Despite the evident benefits of strengthening weather, climate, and water information systems and modernizing NMHSs, considerable efforts are needed just to convince client governments and World Bank teams (at the country and sectoral level) to proceed with project preparation and to commit resources for preparing and implementing complex and often relatively small hydromet modernization activities (see box 6.7 technical insight).

Monitoring Networks

Clients pay the most attention to modernizing monitoring networks and instruments, usually the most expensive part of the program. Modern observation

Box 6.5 Technical Insight: Component 2: Modernization of the Observation Infrastructure and Forecasting

Component 2 aims to (a) modernize the observation networks, communications system, and information and communication technology (ICT) system of the National Meteorological Service (NMS); (b) improve the meteorological (and hydrological, if part of the NMS's responsibility) forecasting system; and (c) refurbish NMS offices and facilities.

This component generally has the following subcomponents and activities:

- Technical modernization of observation networks, which includes
  - Rehabilitating and reequipping meteorological, hydrological, and other networks as required[a]
  - Introducing ground-based remote sensing systems for nowcasting and very-short-range weather forecasting (for example, radar weather surveillance equipment, wind profilers, and lightning detection networks, as appropriate)[b]
  - Including upper-air measurements using radiosondes, although temperature soundings may also be available from commercial aircraft as a part of the AMDAR (Aircraft Meteorological Data Relay) system
  - Strengthening quality control by setting up calibration facilities
- Modernization of the NMS's communication and ICT systems, which includes
  - Introducing new communication equipment that meets World Meteorological Organization (WMO) Information System standards
  - Developing archiving, database management, and digitizing capabilities
  - Providing computers and software to support numerical weather prediction (NWP)[c]
- Improvement of the hydrometeorological forecasting system, which includes introducing modern computer equipment for processing observational data from in situ surface networks, satellites, upper-air stations, radar, and other in situ remote sensing systems (for example, radar, wind profilers, and lightning detection networks)
- Refurbishment of the NMS's offices and facilities because experience shows that facilities that have generally deteriorated and that are in a poor state of repair contribute to poor staff working conditions and are unable to house modern equipment

a. This task may also support aviation, agriculture, or other sector-specific networks, depending on the NMS's responsibility.
b. Generally, all these ground-based remote sensing technologies are complex. If the NMS staff has little or no experience with these technologies, they should be introduced with sufficient long-term technical support. It is also important to weigh alternative technologies that can partly fill the role of these in situ systems, such as satellite-based techniques and systems for nowcasting, which are freely available to WMO Members.
c. In the meantime, or as an alternative, many advanced centers can provide NWPs tailored to a country's specific needs and run them as part of the product suite produced by these centers. Guidance from the WMO and other regional centers is invaluable.

---

methods are typically based on as much automation as possible because automation increases the frequency and reliability of the observations. For example, traditional temperature measurements with mercury-filled thermometers are often recorded only every six hours, whereas automatic stations can report temperatures every 5–10 minutes. During extreme events, when conditions are

Box 6.6 Technical Insight: Component 3: Enhancement of the Service Delivery System

The objective of component 3 is to enhance the service delivery system by creating or strengthening the public weather service (PWS), including developing new information products for vulnerable communities and the main weather-dependent sectors of the country's economy.

Possible main subcomponents are

- Introducing or strengthening the PWS,[a] including services related to disaster risk management, agriculture, the media, civil aviation, health, energy, and water resources. This activity is essential for delivering the benefits of the modernized program. This subcomponent may include the following activities:
  - Developing a service delivery strategy. The PWS will function as the principal interface between the technical provider of products and the users. It should be responsible for developing and implementing standard operating procedures with authorities (such as civil protection) as part of disaster risk management and service quality management.
  - Developing a PWS platform. The PWS platform provides forecasts of the weather's impact on the basis of information available from numerical weather predictions (NWPs), observations, and risk assessments. It requires forecaster workstations for producing decision-support information tailored to each of the PWS sectors. These systems can be turnkey or built to suit.
  - Installing media equipment. Such media equipment enables the National Meteorological Service (NMS) to create broadcast-quality bulletins.
  - Installing computer visualization systems. Such systems should be installed for several user-defined locations (such as civil protection offices, NMS regional offices, and so forth). Each of these systems should be tailored to the specific requirements of the stakeholder.
  - Implementing early-warning system pilots. Pilots are particularly important if the NMS has little or no experience with developing and using warning systems.
- Improving service delivery by introducing mobile telephone–based applications. Although such applications are optional, it is becoming increasingly important to exploit the advances in mobile technology in even the poorest and most vulnerable communities.
- Creating a national climate service. This subcomponent is intended to transform the traditional climatological role of an NMS to a full user-oriented service and to increase opportunities for NMSs to provide relevant climate information to government decision makers and the World Meteorological Organization (WMO). Tasks may include
  - Introducing computer systems to access climate information
  - Developing software to downscale climate forecasts and to translate climate information for decision makers
  - Developing a digital library of all climate-relevant information from all sectors
  - Developing a national framework for climate services (linked to the Global Framework for Climate Services) by engaging all climate-sensitive sectors

a. The PWS provides a model for the delivery of all services, including climate services. Its principal function is to focus on (a) translating and interpreting meteorological and hydrometeorological forecasts into impact forecasts and information and (b) communicating this information to all sectors, including the general public.

> **Box 6.7 Technical Insight: How to Encourage the Preparation of NMHSs' Modernization Projects**
>
> Initial team efforts are usually concentrated on matching the needs for modernizing National Meteorological and Hydrological Services (NMHSs) with opportunities such as identifying champions in the country and within the World Bank and identifying funding commensurable to the task. This process involves
>
> - Seeking government commitment to modernize NMHSs through socioeconomic assessment and building partnerships with project beneficiaries (disaster risk management, agriculture, water resource management, and so forth)
> - Introducing the agenda for hydromet modernization and NMHSs strengthening in the country assistance strategies and country partnership strategies, usually through disaster risk management, climate adaptation, food security, water resource management, and other significant sectoral programs and projects
> - Identifying funding sources for preparation, such as the Global Facility for Disaster Reduction and Recovery (GFDRR) and the Pilot Program for Climate Resilience (PPCR), and for implementation, such as the International Bank for Reconstruction and Development, International Development Association, and PPCR
> - Initially scoping NMHSs modernization activities on the basis of (a) natural risks and vulnerability assessment (types of natural hazards, frequency, and exposure); (b) weather dependence of economy and user needs assessment; (c) evaluation of NMHSs status and high-priority modernization needs; (d) cost-benefit analysis of the potential modernization scenario; and (e) government commitment to sustain proposed modernization.
>
> GFDRR and PPCR resources are proving to be instrumental in supporting these initial stages through technical assistance and economic and sector work.

changing rapidly, this higher-frequency information is particularly useful. But automation does not necessarily mean instruments should be left unattended, given that in extreme environmental conditions instruments need to be inspected.

The transition from entirely manual to even quasi-automation is a difficult, relatively long-term (two to four years), and costly step in all countries, especially countries that rely on unskilled or semiskilled labor and that might be uncertain about future employment (Lynch and Allsopp 2008). NMS management often overlooks the fact that introducing automatic systems requires fundamentally changing data collection routines and operating procedures. This process is costly because it requires extensive staff training, significant technical support, and parallel manual and automatic observations for at least one year at all climatic and other stations with significant historical records.

In many instances, automatic stations are viewed as supplemental to manual stations, and no effort is made to transition the network to fully integrated automatic stations—which is why it is important to engage all staff members, including field observers, so that they fully understand the benefits. Vandalism is also a generic problem in most developing countries, especially where solar panels are used. Experience suggests that where the local community is engaged and can actually access the data directly from the station, vandalism is less of a problem. Over time, the number of observers at the automatic weather station sites should be gradually reduced, and their duties should be transformed from pure observers to technicians, station guards, or community climate extension workers.[1]

Forecasting

Upper-air measurements are particularly important for forecasts (see chapter 3), but many of the least developed countries have stopped taking such measurements because of the cost of the expendables. Continuing to take these measurements requires the government to commit to the additional O&M costs that accompany the new capabilities. In the case of upper-air stations that are critical for global forecasts, donor agencies or the WMO might want to allocate special long-term funding. Because flooding is one of the predominant hazards in many places, most NMHSs need access to tools such as Doppler radar (see box 6.8 technical insight).

Box 6.8 Technical Insight: A Spotlight on Doppler Radar

Nowcasting and very short-range forecasting are vital for hazard warnings of extreme hydrometeorological events (see chapter 4). The most reliable and comprehensive method of tracking atmospheric events in real time is Doppler radar. Doppler radar systems are both complicated and expensive. Considerable technical know-how is needed to maintain them as well as to use the data. However, the benefits are significant in reducing loss of life in the case of flash floods associated with heavy precipitation. In countries with limited technical capacity, such systems will likely be unaffordable or will need to be operated by third parties for a considerable time to prevent system failures. Vendors should be selected with care so that specifications are realistic and can be met. The same applies to instruments such as profilers and to other ground-based remote sensing systems, all of which can be extremely valuable if operated well. Many new techniques are available that use satellite observations and numerical models, which can supplement ground-based systems.

Guidance for all operational matters should be obtained through the World Meteorological Organization (WMO). Reference to WMO basic documents is essential.[a] The Commission for Instruments and Methods of Observation promotes international standardization and compatibility of instruments and methods of observations. The Instrument and Methods of Observation Program sets standards and quality-control procedures and can provide guidance in the use of meteorological instruments and observation methods.

a. "WMO Technical Regulations: Basic Documents No. 2," WMO-No. 49, 2011 edition and later.

A Changing Role for NMHSs

In many least developed countries, NMHSs have a traditional view of the field, which emphasizes collecting observations and producing synoptic weather forecasts. However, the field has changed dramatically—now focusing much more on short-range forecasting of extremes, climate predictions, and delivery of services that meet the specific needs of sectors and stakeholders. Delivering services oriented toward users is a difficult task for many NMHSs, mostly because communicating with users is often an unfamiliar task. But clearly those that provide the desired services have sustainable budgets, and their emphasis should be on training to build a workforce with the right skills.

Program Operating and Maintenance Costs

One of the key factors in defining the project's affordability is assessing the O&M costs of the future system. In developing countries with very limited funding for NMHSs, any modernization effort will generally increase the annual O&M costs. Thus, the government must agree up front to the incremental cost of running the NMS, or the project will be difficult, if not impossible, to sustain beyond the lifetime of the project investment. Experience suggests that the incremental O&M costs are equivalent to approximately 10–15 percent of the investment budget. Many donor-driven projects, which have focused on providing observation equipment, have failed because the NMS or NHS did not have the staff or resources to maintain the new observation capability.

Most developed countries' modernization efforts have also included costs for staff reductions, because employees are often the largest operational expense—in which case, introducing automation is viewed as a cost-reducing strategy. However, this strategy is ineffective in most developing countries, where staffing costs are relatively low. In those countries, relatively few qualified technical personnel are available, and more are usually needed to maintain modern observation and forecasting equipment.

Risk Rating and Mitigation

NMHSs modernization projects are usually high-risk, high-reward projects because they involve considerable institutional changes of technically complex information and communication technology (ICT) and the R&D-based public sector. The risks of such problems can be mitigated by a set of interlinked activities, including the following:

- Building government (ministries of finance, economy, and planning) understanding of the importance of NMSs and NHSs, with the hope of leading to a legally binding commitment fixed in credit or a grant agreement to increase budget support and allocations for O&M costs
- Developing a project design that is more likely to be affordable and implementable by involving the NMS or NHS staff in the process
- Preparing national strategic plans

- Mounting training programs on service provision
- Organizing media workshops
- Building capacity and retaining qualified staff members, including developing additional incentives
- Testing new business development to strengthen sustainability
- Building partnerships with national and international stakeholders, such as twinning arrangements for advanced and developing countries (see box 6.9 technical insight)
- Setting up user and policy committees
- Building the membership of WMO technical commissions.
- Creating a modernization leadership team, composed of NMS and NHS management, the project management unit, and a special modernization team—complemented by specialized technical and implementation support (an integrator or a general consultant)

Procurement Issues

Procurement is complicated owing to the need to buy and install numerous and relatively small packages of various specialized interrelated hardware and software products—often high-capacity computers and communication equipment—and to arrange delivery and installation of equipment in multiple and often

---

Box 6.9 Technical Insight: Types of Twinning Arrangements

It is difficult to realize the full benefit of a modernization program within an institution that does not have a full complement of staff or that lacks training and experience in modern forecasting techniques. National Meteorological and Hydrological Services (NMHSs), however, do not operate in isolation and can benefit from the capacity and capabilities of more advanced services. Most regions of the World Meteorological Organization (WMO) contain both advanced and developing NMHSs. More advantage could be taken of the capacity of the most advanced NMHSs to provide operational support to the weakest. This assistance is provided in some regions, but it is not universal, and it cannot be readily sustained without financial support.

Arrangements for pairing NMHSs are best managed through the WMO, which has the capacity to monitor these activities and where necessary can facilitate the financial support. Ideally, the WMO regional centers would play a leading role in this process because most (but not all) are located within the more advanced NMHSs.

The pairing arrangement would provide operational backup for the forecasters, enabling them to receive guidance on complex weather situations. This guidance could be in the form of bulletins or, in the case of life-threatening situations, through direct video or voice contact, much in the way that a central forecasting office works with regional offices within large countries.

Frequent training is essential to increase and maintain the skills of an NMHS's staff. Such training should be viewed as a continuous process, in which all staff members are involved in a long-term program to improve their skills.

remote locations. A detailed design of networks and systems is expensive and time consuming, and it is often unavailable at the time of appraisal. Consequently, many projects shifted the detailed design of networks to the project implementation phase and included them as the first phase of the integrator or general consultant assignment. The integrator assignment also includes developing most procurement packages (technical specifications and bid documents) and implementing support to NMSs, NHSs, and project implementation units. Hiring an integrator or general consultant is a high priority, and selecting the consultant (usually through quality- and cost-based selection) should be initiated well in advance, given that the project's implementation will strongly depend on the consultant's success.

Two alternatives that the teams considered were ultimately dropped. One involved hiring a group of individual management and technical consultants. This alternative was rejected because NMHSs in developing countries were considered unprepared to provide sufficient coordination and leadership to guide the work of a significant number of individual consultants. The other involved using a turnkey contract for modernization, under which most tasks would be outsourced to a consulting firm. But that alternative was rejected owing to the expected high costs of such an arrangement and limited number of suppliers known to offer such services. The feasibility and the risks of a turnkey approach should be further explored, as well as the feasibility of using supply and installation contracts for the main elements of the observation network.

Results Framework

How should project performance indicators be set? Realistically, it is hard to determine these indicators given that hydromet modernization projects are technically complex and their results are broad and multisectoral. That said, we believe that the indicators need to combine quantitative measures related to forecast accuracy and qualitative measures related to services and benefits:

- *Monitoring and forecast accuracy.* The purpose of this indicator is to measure the technical utility of the new tools—for example, quantity and impact of rain and wind or the accuracy of warnings for specific areas. Standard methods exist for ascertaining the tools' accuracy and should be applied.[2]
- *Public service.* The purpose of this indicator is to measure the impact of the information. Forecasts need to be disseminated in time and evaluated to see if they are understood and can be applied. Evaluation is usually done through public surveys.[3] This measure is probably the most important and useful indication of the utility and success of the NMHSs. It should be considered a high priority and a high-level measure of the program delivery objectives.
- *Social and economic benefits.* The purpose of this indicator is to measure the social and economic benefit of NMHSs products and services. No rigorous or standard methodology has been applied. However, the WMO is beginning to develop an inventory of good practices along with examples of

decision support tools.[4] Good practices are needed to establish a baseline against which to measure the progress of the NMS or NHS modernization efforts.
- *Human resources.* The NMS or NHS depends on the competence of its staff. Modernization projects must include education and training that is substantive and specific to modernization efforts, ranging from forecaster improvement to specialists in delivering services to each client sector of the NMS or NHS.

These issues may be illustrated using the recent example of the results framework being developed for a World Bank project in Nepal, which is considered the fourth most vulnerable country in the world to climate and extreme events (see box 6.10 technical insight).

Box 6.10 Technical Insight: Nepal's Results Framework

The World Bank is assisting Nepal in increasing its resilience to climate change by strengthening the Department of Hydrology and Meteorology (DHM) to provide essential weather, climate, and water services. The goals of the Pilot Program for Climate Resilience (PPCR) Project on Building Resilience to Climate Related Hazards in Nepal are to (a) strengthen infrastructure; (b) modernize observation and forecasting systems; and (c) introduce public weather, climate, and hydrological services. The results framework has three main project development objective indicators:

- *Increased sustainability of DHM operations.* The percentage of public funds allocated for essential operational needs would be a unit measure for the first indicator. Although the availability of public funding is not an ideal measure of institutional sustainability, it can be a proxy at this stage. Another difficulty is that no formal procedure exists to estimate essential operational needs for the DHM. For instance, the baseline value of this indicator (40 percent) is based on a DHM assessment and discussions with the World Bank project preparation team.
- *Increased accuracy and timeliness of weather forecasts.* This indicator—which is well established and widely used in advanced NMHSs—depends on forecasters' skills, information and communication technology (ICT), and observation infrastructure. But in many developing countries, it is difficult to apply because National Meteorological Services (NMSs) often either do not measure the accuracy of forecasts or measure them using homemade procedures. In the Nepal PPCR project, it was suggested that a forecast verification system in DHM operations be introduced against which project-related improvements could later be measured. Thus, the proposed target of forecast accuracy is speculative in the project preparation stage and will be adjusted on the basis of the results of verification.

box continues next page

Box 6.10 Technical Insight: Nepal's Results Framework *(continued)*

- *Increased satisfaction of users with DHM services.* Increased user satisfaction is the most integrated indicator, but it is difficult to introduce and relatively costly to measure. A composite satisfaction index expressed as a percentage, with 100 percent showing complete user satisfaction, is suggested for the unit measure. There is no history of using such indexes in most developing countries. So the methodology for using this index in a particular country will have to be developed and a baseline established. In most developed countries that use similar indexes, the satisfaction level of a mature service is expected to exceed 85 or 90 percent. In a developing country, a satisfactory level may be much lower (65 percent) at the end of five years; however, an upward trend should be noted and expected to continue.

Safeguards

Environmental and social risks of NMHSs modernization projects are typically small. Moreover, the hydromet projects usually provide environmental benefits, given that they support lower risks associated with floods, drought and fire, winds, extreme weather events, and even industrial accidents. Project activities are usually implemented within available hydromet sites and involve a minor installation of observation equipment with no or minimal environmental disturbance. Specific attention should be given to the safe handling and disposal of equipment containing mercury (for example, thermometers). Procuring larger equipment (such as Doppler radar) or constructing buildings will require a more detailed evaluation of the public health and environmental safety aspects, which should be carefully documented in the environmental and social management framework. A potential staff reduction during a large-scale replacement of manual observations by automatic instruments should be carefully evaluated within the framework's scope and will require extensive consultations with key stakeholders. That said, so far, no hydromet modernization projects in developing countries have triggered any incremental staff reductions.

In sum, the capacity to cope with extreme hydrometeorological events depends on timely action in response to accurate and useful weather, climate, and hydrological forecasts. The most important life-saving actions usually occur in the few hours preceding the event's impact. Predicting with certainty the intensity and location of rainfall is often vital to keeping people out of harm from a flash flood, for example. Traditional synoptic forecasts, which depend on six hours of data, are not adequate for this task. Much higher temporal and spatial observations are needed from automated in situ observations and, where possible, ground-based remote sensing (such as Doppler radar). In most instances, NMHSs need a complete overhaul of their operations—operating procedures, legal frameworks, observation networks, forecasting systems, and service delivery. But each of them is interdependent. The purpose of upgrading the observation networks is lost if the new data cannot or do not improve the quality of the forecasts. And none of it matters if services fail to reduce loss of life and livelihoods. Thus, a holistic approach to the entire meteorological or hydrological service is essential.

A frequent problem for modernization programs is the incremental O&M costs. In fact, O&M costs are often the primary reason automated observation networks fail after only a few years in operation. But if they fail, forecasts fail, and then confidence is lost—and once again, the system spirals into decline. There is no simple solution if governments are unwilling to accept this responsibility. In many countries where the status of NMHSs is poor, modernization is extremely challenging. In those cases, governments should seek the assistance of the World Bank, the WMO, and other development partners. At the World Bank, more experience is being gained with each new project, and we will continue to refine and share best practices.

Notes

1. Creating a cadre of climate extension workers from existing observers could help improve the community's response to hazardous weather and increase local knowledge on climate resilience.
2. This web page, which is maintained by Australia's Bureau of Meteorology, provides access to the latest information on forecast verification: http://www.cawcr.gov.au/projects/verification.
3. This web page, which is maintained by the WMO, provides examples of surveys that can be applied by any NMS: http://www.wmo.int/pages/prog/amp/pwsp/surveys.htm.
4. See the WMO's web page on the socioeconomic benefits of weather, climate, and water services at http://www.wmo.int/pages/prog/amp/pwsp/SocioEconomicMainPage.htm.

References

Chomitz, Kenneth, Dinara Akhmetova, and Stephen Hutton. 2012. *Adapting to Climate Change: Assessing the World Bank Group Experience, Phase III*. Washington, DC: World Bank.

Hancock, Lucy, and Vladimir Tsirkunov. 2013. "Strengthening Hydrometeorological Services: The World Bank Portfolio Review (1996 to 2012)." World Bank, Washington, DC.

Lynch, Daryll, and Terry Allsopp. 2008. "Automated Versus Manual Surface Meteorological Observations: Decision Factor." World Meteorological Organization, Geneva, Switzerland.

NRC (National Research Council). 2011. *The National Weather Service Modernization and Associated Restructuring: A Retrospective Assessment*. Washington, DC: Nation Academies Press.

———. 2012. *Weather Services for the Nation: Becoming Second to None*. Washington, DC: National Academies Press.

Pilon, Paul J., and Kidane Asefa. 2011. *Comprehensive Review of the World Hydrological Cycle Observing System*. World Meteorological Organization, Geneva, Switzerland.

Zaitsev, Alexander, and A. Yagupov. 2010. *Swiss Support to Hydro-meteorological Services in the Aral Sea Basin*. Completion mission report, Swiss Agency for Development and Cooperation, Bern.

Environmental Benefits Statement

The World Bank is committed to reducing its environmental footprint. In support of this commitment, the Office of the Publisher leverages electronic publishing options and print-on-demand technology, which is located in regional hubs worldwide. Together, these initiatives enable print runs to be lowered and shipping distances decreased, resulting in reduced paper consumption, chemical use, greenhouse gas emissions, and waste.

The Office of the Publisher follows the recommended standards for paper use set by the Green Press Initiative. Whenever possible, books are printed on 50% to 100% postconsumer recycled paper, and at least 50% of the fiber in our book paper is either unbleached or bleached using Totally Chlorine Free (TCF), Processed Chlorine Free (PCF), or Enhanced Elemental Chlorine Free (EECF) processes.

More information about the Bank's environmental philosophy can be found at http://crinfo.worldbank.org/crinfo/environmental_responsibility/index.html.